Sustainable Development in the Sahel

Proceedings of the 4th Sahel Workshop, 6–8 January 1992

editors

A. M. Lykke, K. Tybirk & A. Jørgensen

AAU REPORTS 29

Dept. of Systematic Botany, Aarhus University, 1992

Contents

Anne Mette Lykke, Knud Tybirk and **Agnete Jørgensen** (Eds.) — Introduction (p. iii)

Kwesi Kwaa Prah — Sustainability of Production Systems and Society in Arid and Semi-arid Africa (p.1)

Fantu Cheru — Debt Crisis, National Agricultural Policies and Sustainable Development in the Sahel (p.13)

Mike Speirs — Development of Indigenous Integrated Farming Systems in the Sahel (p. 23)

Henk Breman — The Sustainability Concept in Relation to Rural Development in the Sahel: Offering Stones for Bread (p. 37)

Anette Reenberg and **Kjeld Rasmussen** — Problems of Defining and Evaluating Sustainability of Agricultural Systems (p.53)

Poul A. Sihm — Pastoral Associations and Natural Resource Management (p. 67)

Sofus Christiansen — The UNCED Process and the Sahel (p. 73)

Moussa Seck, Phillipe Engelhard and **Taoufik Ben Abdallah** — Beyond the Agricultural Crisis in the Sahel (p. 79)

Søren Leth-Nissen — Sustainable Development: just Another Pet? (p. 85)

Søren Skou Rasmussen — The Nile, the Desert and the People — towards a Cultural Strategy for Environmental Action (p. 89)

Anders Michelsen — The Role of Nodulating Bacteria and Mycorrhizal Fungi for Growth and Establishment of Trees in Ethiopia and Somalia (p. 97)

Anders Michelsen, Ib Friis and **Lisanework Nigatu** — Aspects of Nutrient Cycling in Three Plantations and a Natural Forest in Ethiopia (p. 99)

Henning Høgh Jensen — Agriculture and Sustainability in a Danish NGO Context (p. 103)

Jens Weise Olesen — Famine and Colonial Administration (p. 107)

Mike Speirs — Notes from Concluding Session and Evaluation (p. 119)

Summaries of Participants' Projects (p. 121)

Address List of Participants (p. 129)

Editors

Anne Mette Lykke. *Born 1963. Since 1991 Ph.D.-student at Department of Systematic Botany at the Institute of Biological Sciences, Aarhus University. Working on a project entitled: Sustainable Use of Natural Vegetation in the Sine Saloum Region in Senegal. Address: 68, Nordlandsvej, DK-8000 Aarhus, Denmark.*

Knud Tybirk. *Born 1960. Cand. scient. Aarhus University 1988. Ph.D. at Department of Systematic Botany at the Institute of Biological Sciences, Aarhus University 1992. From 1988–91 research associate financed by Danish Research Council for Development Studies (Danida) on a project entitled: Survival Strategies of Sahelian Woody Legumes in Relation to Management. Present address: c/o FAO, Apartado 1721-0190, Quito, Ecuador.*

Agnete Jørgensen. *Born 1959. M. Sc.-student at Department of Systematic Botany at the Institute of Biological Sciences, Aarhus University. Working on a project entitled: Reproduction of* Adansonia digitata. *Address: 68, Nordlandsvej, DK-8000 Aarhus, Denmark.*

Introduction

This volume contains papers presented at and inspired by the *4th Sahel Workshop* held in Gilleleje, Denmark, 6–8 January 1992. A Summary of workshop participants' projects is included to present aspects of their work.

The Sahel Workshop is an interdisciplinary approach of researchers, students, donors, private firms and Non-Governmental Organizations working in the Sahel, or in similar semi-arid areas of Africa, to discuss the problems of the Sahel.

This year the theme of the workshop was the concept of *Sustainable Development*, as coined by the Brundtland-report *Our Common Future* in 1987. The presentations and discussions showed how difficult, and at the same time how stimulating, it was to discuss sustainable development and the many different definitions of the concept. The workshop covered economical, social, political, agricultural and ecological sustainability in the Sahel. The wide scope of approaches made it obvious that the concept of sustainable development needs more research in order to understand all its ramifications in the Sahel and semi-arid Africa in general.

We thank Neil Gale for linguistic corrections and Anni Sloth and Kirsten Tind for technical assistance. The workshop was financed by Danida (Journal no. 104.P.3.Sahel).

A. M. Lykke
K. Tybirk
A. Jørgensen

Aarhus, May 1992.

Sustainability of Production Systems and Society in Arid and Semi-arid Africa

Kwesi Kwaa Prah
University of Cape Town, South Africa

Introduction

Few issues affecting the human species and the prospects for its future on this globe have engaged the minds and consciences of people in all corners of the contemporary world as much as the problem of the environment. From tropical rain forests to areas of extreme aridity, concerned people are increasingly, and with even stronger voices, drawing attention to the gradual debilitation of the environment and the ceaseless destabilization of its balance and fragility. Its ability to hold and carry populations and production systems for society is being undermined by a history of carelessness, rampant and irreplaceable pillage of its resources, and uncontrolled stress.

The development idea has in the past shown little respect for the environment. Resources were extracted and exploited with little thought for their possible replacement or the polluting consequences of their processing, as long as they contributed to the creation of wealth and satisfied consumer demands in terms of the goods produced. The environment was indeed treated as if it had limitless resources.

It has been the gradual awakening of society to the dangers of the unbridled and irreplaceable exploitation of environmental resources that triggered the conception of sustainable development on us. The interest in environmentally-related issues has spawned in its wake a tendency for some scholars to look to the environment in every issue. Environmental issues are more easily funded by donors, research, and aid agencies. The anthropologist Palsson (1990a) is obviously sensitive to these issues when he discusses the "deterministic aspect of cultural ecology" and observes that "the discussion of arid environments involves fundamental theoretical issues, apart from that of the observation and measurement of environmental interactions. No one would deny that the ecosystem is important for human life, but social theorists debate just how to incorporate the ecological dimension into anthropological analysis."

Pastoralist societies in arid and semi-arid Africa, particularly in the Sahel, have performed poorly in the overall development endeavour.

Where state intervention in production has been pushed, it has only marginally augmented the productive capacity of these countries. A good example in this respect is provided by the case of the Operation Riz-Segou in Mali. Initiated as an attempt to expand rice production in Mali, the para-statal agency soon created its bureaucratic control structure in order to reduce and hamster production incentives and stifle smallholder control. It has over the years reinforced and created bases of power that serve economic and political interests alien to most smallholders. It has been suggested that, under these conditions, foreign aid has tended to strengthen the influence of bureaucrats and become negative and counter-productive to long-term agricultural development (Bingen 1985).

Sustainable development is essentially a concept which tries to relate economic growth with ecological concerns. It can be understood ideally, as an economy using only renewable resources, and those only at the rate at which they are replenished. If this is to happen, then the increasingly identifiable sources of deepening economic crisis and poverty in (for our purposes) the arid and semi-arid lands of Africa need to be addressed in ways which permit the sustainability of development.

The theme paper for this conference makes a number of important points with regard to the concept of an environmentally sound and sustainable development. In the context of "a process that did not result in a degradation of the natural resources including the soils, the vegetation, and the water resources," we note the fact that "the absence of land degradation is no guarantee... that the socio-cultural conditions evolve satisfactorily or that the production is economically viable." Obviously, if arid area dwellers at the subsistence level cannot make ends meet and keep body and soul together at a sustainable level, there is little chance that the pressures they put on the environment would be guided and mitigated by any concerns for the protection of the environmental resources. My argument is that if life is economically sustainable the chances of the environment being kept at sustainable developmental levels is more assured. I also mean that development projects which upgrade the environment while providing a livelihood for arid area dwellers should be particularly favoured. A good example of this is provided by the Senegal River Project which apart from providing irrigated crops for rural and urban centres is providing the opportunity for the "genesis of a man-created landscape replacing the natural deserts and grasslands." As Thiel (1991) has noted, "the water now trapped by the dams which no longer flows wastefully to the sea in the rainy season could also be used for large-scale afforestation. Along the roads parallel to the Senegal River many of the trees have been cut down but few new trees have been planted. This is a task for ecological development aid, individual cooperatives have already started to grow and plant saplings, but it will take years before these yield any benefit. Afforestation entails enormous sacrifice for people living at subsistence level."

It is noteworthy that Scandinavian scholarships on the problems of arid and semi-arid lands represents today one of the most relatively concentrated, significant, and distinctive emerging traditions in this area of concern. Over the past decade and a half, a body of literature has emerged from the activity of Nordic-based and Nordic-sponsored studies on arid and semi-arid areas in Africa. Collectively, they provide a wealth of insights which can be summarized and consolidated to provide a reference framework for the design and implementation of programmes and projects with sensitivity for developmental sustainability. In this paper, while drawing on my own observations and experience from a variety of African countries and other sources, my discussion is situated against the backdrop of Nordic scholarships on the problems of the arid and semi-arid areas of Africa[1].

The arid areas of the tropical world have average rainfall figures of between 100 and 400 millimetres per year. The areas receiving 350-400 millimetres of rainfall are more or less those areas which lie at the threshold of not requiring irrigation for cultivation (Stryker 1989). Over half of the African continent is either arid, semi-arid or extremely arid. The unreliability of precipitation and the consequent drought-proness often makes drought and rain major institutional symbols in the thinking and mythology of societies situated in arid and semi-arid areas. Rain-makers and rain-making become central institutions in the collective lives of people. Among the Tswana of Botswana, South Africa and Namibia, *Pula* is at one and the same time a word for rain, all that is bountiful, blessing, honour, and peace (Prah 1980). A similar point has been made by Dahl and Mergessa (1990) regarding the Borana of the Ethiopia-Kenya borderlands. They indicate that ideas about fertility and descent are linked with collective understandings of the blessings of wells, and wells are closely attached to the basic concepts of Boran identity. For the Borana water is life. "Water is more than a physiological necessity: it is a central ontological concern." O'Leary (1990) makes the point that the collective memory of the nomadic pastoralists is crystallize in their event calendars. In his study of the Rendille and Gabra of Marsabit District, Kenya, he found that there was indeed a close correspondence, with few exceptions, between the dating of droughts based on climatic data, official reports, and the local event calendars. Ndagala (1990) has recorded the great value placed on vegetation, particularly grass, by the Maasai of Tanzania. According to the Maasai, cattle and grass go hand in

---

1 A representative and cross-sectional example of such studies is what has, over the years, been produced under the auspices of the Scandinavian Institute of African Studies (SIAS) in Uppsala. Of particular significance in this respect has been the work emanating from the project on "Human Life in Arid Lands" coordinated by Hjort af Ornas and Salih (1989). I have used material from various SIAS publications as the basis for this paper. My remarks are largely attributable to sub-Sahan Africa. In a statement on pastoralism and the development challenge, Poulsen (1990) has drawn attention to the useful work of the "Sahel Workshop."

hand and were provided by God (Engai). The scarcity of grass on account of drought leads to ruin. Social, economic and cultural patterns reflect historically and traditionally adaptations to suit the nature of the environment and its resources. As have been rightly observed by Bovin and Manger (1990) resource scarcity, poor and unpredictable rainfall, and relatively unproductive soils make production in these areas a risky endeavour and have undoubtedly influenced societies in these areas, over the centuries, to adopt flexible habitational and production strategies. The precariousness and inability for societies in these areas to support themselves have certainly on the whole increased dramatically over the past two decades. External food aid has almost become a permanent feature of many of these societies. This is, however, of longer standing than many observers realize, dating back to the colonial period. In his memoirs, Sir Donald Cameron (1939), the British administrator had this to say, "I always maintained when I was in Tanganyika, that it was a somewhat cruel climate for the agriculturalist (except in the case of specialised crops like sisal and coffee), and I know of no reason for modifying that statement now. I saw six harvests in the Territory, and one only of those was fairly good; even in that comparatively favourable year there were parts of the country in which the harvest had failed and we had to feed the people. And none of these failures were at all attributable to locust invasion; they were all due to lack of rain at the time it was needed and to no other cause." Cameron follows these remarks up by pointing out that he had read, with consternation reports from the districts indicating that under conditions of crop failure, in some localities people had gone out into the woods "as a matter of course to dig up roots and endeavour to find other precarious means of keeping themselves from starvation." What he obviously did not fully understand was the fact that such dietary changes represented set patterns of adaptations which such societies had over decades evolved, not as fixed cultural habits, but as dynamic adaptations to a changing environment and available resources. In an earlier paper I have drawn attention to how such dietary changes are triggered among the Sarwa (San) of Botswana (Prah 1978).

Social organization and production systems

In the Sahel and other arid or semi-arid areas of Africa, traditionally four main types of production systems have been prevalent. They overlap in many ways and may exist almost conjointly in some societies. The most socio-structurally simple are hunter-gatherer communities located largely for our purposes in southern Africa. Pastoralist and agro-pastoralist systems are the most common types of production systems found in these areas. The rarer or less common system for arid lands is arable cropping and consolidated sedentism.

Layton *et al.* (1991) depart from what in the past has been very orthodox in accepting the transition from hunting and gathering to herding or cultivation as a more or less evolutionary and historical progression. The authors advance an "optimal foraging theory", which suggests the usefulness of treating hunting, gathering, herding, and cultivation as alternative strategies which are singly or in combination, suitable for the exploitation of specific social or natural environments. Essentially in their view, hunter-gatherers rely on a range of foods with differential yields in terms of the labour power and time invested. The availability of those providing the best yields vary over space and time. Thus according to this theory, "...setting out across a familiar landscape and encountering a number of potential sources of food and predicts which will be judged worth stopping to exploit and which will be ignored in the expectation that something more rewarding will be found later. The simplest measure of reward for stopping to exploit a resource is its yield in calories per kilo. The costs include the rate at which the resource is encountered, pursuit time (if it is mobile), and preparation time (*e.g.,* butchering a carcass, cracking nuts, or winnowing and grinding seeds). Calculations of the relative costs and benefits of exploiting different foods yield a rank order of food sources which can be tested against observed preferences."

According to Layton *et al.* (1991) reversals occur in the production system as adaptive options change. In diverse ecosystems, hunting and gathering may exist side by side, or in symbiosis with pastoralism and cultivation. Loss of husbanded resources may favour hunting and gathering, while the degradation of the environment and the pressures of population growth tend to render intensified husbandry irreversible. The logic of the above argument is suggestive, facile, and attractive, but falls flat on the face in its inability to deal with the problem of historicity of production systems. The rejection of what I would describe as "palaeo-evolutionism", or what has been defined by Ingold (1980) as the claim that the transition from hunter-gathering to pastoralism and or sedentary cultivation "in some absolute sense constitutes progress" is possibly too overwhelmingly accepted to invite serious controversy. The bath water, however, must not be thrown out with the baby. Fear and apprehension about the pitfalls of crude historicism should not constitute grounds for equally crude ahistoricism. Layton *et al.* (1991) specifically point to a number of factors which in their view may render a shift from hunting and gathering to intensive husbandry. These include in their estimation climatic change, technological innovation, "the elaboration of social networks," and the appearance of new varieties of plant and animal life which facilitate and enhance intensive husbandry either through local mutation or cultural diffusion. They rightly argue that none of these factors necessarily favour or prompt a unidirectional transition. The important fact of the matter, however, is that in the broad historical sweep, societies individually or collectively do not unwind or scale down in their complexity and unlearn their cultures (I am using the concept of culture here to encapsulate the totality of the human historical product of a given society or societies, without necessarily implying determinism). The value of Ingolds observation that "pastoralism can only be distinguished from hunting in terms of its social relations (of divided access to animals as property) which in turn specify the conditions of adaptation" deserves closer attention. A more structured, less reified and historically responsible formulation would need to be advanced to explain transitional processes of production systems.

My observations on the African ethno-ecological map would suggest that in arid and semi-arid areas socio-structural complexity and stratificational patterns tend to be largely confined to cultivators and pastoralists as opposed to hunter-gatherers. Some pastoralist groups are highly stratified with elaborate pre-colonial state structures, while others have simpler forms of social organization. Sometimes, within one ethnic group or cluster of related ethnicities such variation is evident. A good example of this would be the Herero and the less socio-organizationally complex Himba of Namibia. Another example is provided by Bovin (1990) in her discussion of the Fulani of the Sahel. She points out that the Fulani inhabit fairly different ecological regions of the Sahelian and savanna areas of West Africa. They are indeed one of the most ubiquitous ethno-linguistic groups in West Africa, with elements in Nigeria, the Cameroons, Chad, Niger, Mali, Senegal, Guinea, Burkina Faso and Mauritania. Their adaptations with respect to the environment vary from nomadic pastoralism to semi-nomadic or semi-sedentary agro-pastoralism to more or less almost complete sedentary cultivation. Bovin points out that these different adaptations are also in part related to the historical developments surrounding the spread of Islam in the region and the emergence of pre-colonial Fulani states throughout the region. The two groups which Bovin discusses in detail in her recent study are the Fulbe Liptaako of northern Burkina, an agro-pastoralist stratified class society which consists of residual elements of the former Emirate of Liptaako, and the Wodaabe, a nomadic pastoralist group moving across parts of Chad, Cameroun, Niger and Nigeria. They have institutionalised a social organization with lineages, clans, and inhabit camps, and not essentially villages or towns.

Cultivators are not necessarily more complex in their social organization than pastoralists. In areas where such groups coexist a range of other factors such as technology, political system, population size, social networks, economic, and market conditions are crucial in determining dominance, power, and influence. One would need to point out, however, that in as far as food production and the attainment of calorific minima are concerned, cultivators and pastoralist differ from hunter-gatherers on their reliance on a more limited range of food sources. They obviate the dangers of starvation through the application of intense husbandry and production methods of preferable resources.

Among the Karamojong, Turkana and Suk of Uganda and Kenya pastoralist mobility and the risk of stock and human loss through raiding and disease promote a communalistic and egalitarian ethos. The Hausa-Fulani have over the centuries developed highly hierarchized institutions stimulated by the returns which long-distance trade and barter bring.

The development challenge

Outside the ambit of the question of the development challenge *per se,* it is possible to note trajectories of transformations affecting societies in arid and semi-arid areas in contemporary Africa. These transformations are in part beyond the control of the active local participants in as far as some of them, possibly the most significant, are inspired by forces and interests lying outside the immediate social, cultural, and economic worlds of these societies. Some of these influences spring from implications arising out of the transformations in the international economic system, the north-south structure and the process of integrating periphery societies into the international economic order.

Other influences radiate from local structures created by the post-colonial state particularly with reference to covert and overt political coercion and economic stimuli. Examples of this are provided in the work of Salih (1989), Ahmed (1989), and Christiansson and Tobisson (1989). Conflict and war which in recent years have become so endemic in sub-Saharan Africa are also important factors. Semait (1989) has suggested in a study of Eritrea, Ogaden, and Tigray that there is "spatial coincidence between ecological stress and political conflict....the precise cause and effect relationship cannot be established without carrying out further research." Among the direct consequences of armed conflicts on the regions of Africa under ecological stress, Cervenka (1989) lists the suspension of development projects, the destruction of cattle herds (the insurance for the survival of pastoral people), the erosion of morality, damage to culture, the deterioration of the status of women, and the violation of human rights. Nnoli (1989) observes that, "unless there is peace, development is not possible ... unless there is development, peace is not possible and if it exists it will not last."

Technological innovations, especially those that are not integrated into the economic complex of pastoralists, but which are engendered largely by the penetration of the cash nexus may provide scope for the elevation of the Gross Domestic Product of a given country. But there is no way of assuring that even when development projects are situated in the middle of arid lands, the ultimate and real benefits are reflected in the life conditions of the pastoralist, agro-pastoralist or hunter-gatherer. Note should be made of the useful questions raised by Palsson (1990b) regarding expanded and improved technologies in the form of larger fishing boats for Cape Verdean fishermen. In response to a World Bank suggestion that, "if all small-scale fishermen were to fish from medium size boats their aggregate annual catch might be some 25,000 tons," Palsson remarks that while this may well be the case a number of questions need to be addressed, "are such prospects realistic for Cape Verde, or any Third World country for that matter? Would such efforts violate the reproductive potential of the fishing stocks? And what sort of social repercussions would such a restructuring of the industry entail."

Kituyi's (1990) study[2] on the socio-economic transformations affecting the pastoral Maasai reveals that a whole range of novel patterns of existence are affecting the pastoral Maasai and unhinging them from their traditional moorings as they "become Kenyans." If they were more involved and sensitized to the nature and the direction of these forces of transformation, their ability to deal with developmental issues more creatively would be enhanced. Across the border in Tanzania, Århem (1985) discovered that, the Maasai of the Ngorogoro Conservation Area "feel subject to too many rules and restrictions imposed upon them by the authority." Population increase and pressures constitute another important factor affecting the development question. It has also a profound impact of the environment and its ability to absorb stress. Krokfors (1989) has made the very interesting observation that essentially the politicisation of land-based resources is a direct reflection of a given society's political structure. She brings the political factor into useful interface with the population question and shows that environmental stress, land degradation, and over population relate to the political pattern of resource allocation. Myers (1989), in line with conventional wisdom, is inherently gloomy about Africa's prospects particularly in the continent's possible inability to provide the food requirements of its projected population increase. The security implications at a regional level for the control of river systems is interesting and need interest and attention. It needs, however, to be pointed out that population growth does not in my thinking necessarily need to have negative consequences on the environment or inhibit economic growth. Indeed, it is conceivable that under certain conditions it can support economic growth. But such conditions have hardly ever been engendered in sub-Saharan Africa.

Economic growth and development constitute to my mind the crucial factor for better and sounder environmental preservation in Africa. Poverty and the inability to sustain life in manageable fashion promotes a careless and stressful attitude towards the environment. It

---

2 Kituyi shows, however, in this text that while major social transformations are taking place in Maasai society, certain older institutions remain thus far absolute and unchanged. I would argue that greater involvement in active and real decision-making would assure less trauma in the face of social change and enlist the support of such pastoralists to a better and more socially meaningful and constructive fashion. If sustainable development is to work in such areas, the direct involvement of the pastoralist in decision-making is invaluable.

also makes society more liable to security breakdowns, forced migrations, and conflicts which stress and provoke the vulnerability of the environment. Sustainable development means also that there is need for computation of the costs of protecting the environment.

The maintenance of the quality of the environment through the protection of its resources as part of the factors of production in turn serves as a threshold factor for the development and maintenance of the quality of life. Economic thought which is ecologically sensitive needs to expand the concept of profitability beyond considerations of running costs and reserves for repair and maintenance. Conceptually, environmental costs need to be included in economic profitability calculations.

The emergence of the concept of sustainable development represents a milestone in the progress of mankind to civilize responses to the environment as an inherited resource which cannot be replaced if destroyed, and which maintains such a basic relationship to all life (including human) that its destruction or degradation implies directly the destruction and degradation of the species.

From the findings and papers that have over the years appeared in the forum of the "Sahel Workshop," it is possible to identify a number of interacting factors which affect the quality of the environment in arid Africa, and which provide clues as to the issues to be attended. I have raised some of these issues in this discussion. A careful distillation of the Sahel Workshop's past work should provide guide-lines for designing projects in arid Africa, which would be environmentally sound and developmentally sustainable.

Another point which needs to be increasingly made is the fact that such programmes and projects need to involve the human objectives in the designing and implementation of the programmes and projects, at levels in which inputs of whatever sort are ultimately organically integrated into customary usages by pastoralists and agro-pastoralists. The democratization of the process of creating sustainable development projects is to my mind the ultimate guarantee for the success and indigenization of sustainable development thinking. Sustainable development is only possible if the pastoralist and agro-pastoralist are in charge of their own destinies.

Literature cited

Ahmed, A. G. M. 1989. Ecological degradation in the Sahel: the political dimension. — Pp. 89–101 in: Hjort af Ornas, A. and Salih, M. A. M. (Eds.), Ecology and politics. Environmental stress and security in Africa. The Scandinavian Institute of African Studies, Uppsala.

Bingen. R. J. 1985. Food production and rural development in the Sahel. Lessons from Mali's Operation Riz-Segou. Boulder.

Bovin, M. 1990. Nomads of the drought: Fulbe and Wodaabee nomads between power and marginalization in the Sahel of Burkina Faso and Niger Republic. — Pp. 29–59 in: Bovin, M. and Manger, L. (Eds.), Adaptative strategies in African

arid lands. The Scandinavian Institute of African Studies, Uppsala.
Bovin, M. and Manger, L. 1990. Introduction. — Pp. 9–29 in: Bovin, M. and Manger, L. (Eds.), Adaptative strategies in African arid lands. The Scandinavian Institute of African Studies, Uppsala.
Cameron, D. 1939. My Tanganyika service and some Nigeria. — Unwin Brothers, London.
Cervenka, Z. 1989. The relationship between armed conflicts and environmental degradation in Africa. — Pp. 25–37 in: Hjort af Ornas, A. and Salih, M. A. M. (Eds.), Ecology and politics. Environmental stress and security in Africa. The Scandinavian Institute of African Studies, Uppsala.
Christiansson, C. and Tobisson, E. 1989. Environmental degradation as a consequence of socio-political conflicts in Eastern Mara Region. — Pp. 51–67 in: Hjort af Ornas, A. and Salih, M. A. M. (Eds.), Ecology and politics. Environmental stress and security in Africa. The Scandinavian Institute of African Studies, Uppsala.
Dahl, G. and Megerssa, G. 1990. The sources of life: Boran concepts of wells and water. — Pp. 21–39 in: Palsson, G. (Ed.), From water to world-making. The Scandinavian Institute of African Studies, Uppsala.
Hjort af Ornas, A. and Salih, M. A. M. 1989 Ecology and politics. Environmental stress and security in Africa. The Scandinavian Institute of African Studies, Uppsala.
Ingold, T. 1980. Hunters, pastoralists, and ranchers. — Cambridge University Press, Cambridge.
Ingold, T. 1991. Comment on; Layton, R., Foley, R. and Williams, E. The transition between hunting and gathering and the specialized husbandry of resources. A socio-ecological approach. — Current Anthropology 32: 255–275.
Kituyi, M. 1990. Becoming Kenyans. Socio-economic transformation of the pastoral Maasai. Nairobi.
Krokfors, C. 1989. Population and land degradation. — Pp. 197–211 in: Hjort af Ornas, A. and Salih, M. A. M. (Eds.), Ecology and politics. Environmental stress and security in Africa. The Scandinavian Institute of African Studies, Uppsala.
Layton, R., Foley, R. and Williams, E. 1991. The transition between hunting and gathering and the specialized husbandry of resources. A socio-ecological approach. — Current Anthropology 32: 255–275.
Myers, N. 1989. Population growth, environmental decline and security issues in Sub-Saharan Africa. — Pp. 211–233 in: Hjort af Ornas, A. and Salih, M. A. M. (Eds.), Ecology and politics. Environmental stress and security in Africa. The Scandinavian Institute of African Studies, Uppsala.
Ndagala, D. K. 1990. Pastoral territoriality and land degradation in Tanzania. — Pp. 175–189 in: Palsson, G. (Ed.), From water to World-making. The Scandinavian Institute of African Studies, Uppsala.
Nnoli, O. 1989. Desertification, refugees and regional conflicts in West Africa. — Pp. 169–181 in: Hjort af Ornas, A. and Salih, M. A. M. (Eds.), Ecology and politics. Environmental stress and security in Africa. The Scandinavian Institute of African Studies, Uppsala.
O'Leary, M. F. 1990. Drought and change amongst nomadic pastoralist: the case of the Rendille and Gabra. — Pp. 151–175 in: Palsson, G. (Ed.), From water to World-making. The Scandinavian Institute of African Studies, Uppsala.
Palsson, G. 1990a. Introduction. — Pp. 7–21 in: Palsson, G. (Ed.), From water to World-making. The Scandinavian Institute of African Studies, Uppsala.
Palsson, G. 1990b. Cultural models in Cape Verdean fishing. — Pp.93–109 in: Palsson, G. (Ed.), From water to World-making. The Scandinavian Institute of African Studies, Uppsala.
Poulsen, E. 1990. The changing patterns of pastoral production in Somali society. — Pp. 135–151 in: Palsson, G. (Ed.), From water to world-making. The Scandinavian Institute of African Studies, Uppsala.
Prah, K. K. 1978. Some sociological aspects of drought. — in: Hinchey, T. (Ed.),

Botswana Society Conference Proceedings.

Prah., K. K. 1980. The state and traditional responses to drought among the Tswana. — in: Kiros, F. (Ed.), EASSCG Conference Proceedings, Addis Ababa.

Salih, M. A. M. 1989. Ecological stress, political coercion and the limits of state intervention, Sudan. — Pp. 101–117 in: Hjort af Ornas, A. and Salih, M. A. M. (Eds.), Ecology and politics. Environmental stress and security in Africa. The Scandinavian Institute of African Studies, Uppsala.

Semait, B. W. 1989. Ecological stress and political conflict in Africa: the case of Ethiopia. — Pp. 37–51 in: Hjort af Ornas, A. and Salih, M. A. M. (Eds.), Ecology and politics. Environmental stress and security in Africa. The Scandinavian Institute of African Studies, Uppsala.

Stryker, J. D. 1989. Technology, human pressure, and ecology in the arid and semi-arid tropics. — in: Leonard, H. J. *et al.* (Eds.), Environment and the poor: development strategies for a common agenda. Overseas Development Council, Washington.

Thiel, R.E. 1991. Senegal: water for the desert? — in: Development Cooperation. DSE.No.4.

Århem, K. 1985. Pastoral man in the garden of Eden. — Uppsala Research Reports in Cultural Anthropology, University of Uppsala, Uppsala.

Debt Crisis, National Agriculturel Policies and Sustainable Development in the Sahel

Fantu Cheru
The American University Washington D.C., U.S.A.

Introduction

The problem of Africa's debts has increasingly attracted the attention of African and international policy makers in recent years, and a variety of remedial actions have been initiated. Unfortunately, these actions are being taken without due consideration to environmentally and socially sustainable forms of agricultural production patterns. The policies being adopted are based on wrong assumptions and treat the debt crisis as a separate activity from the task formulating sustainable agricultural development policies.

The management of the debt crisis — *a la* structural adjustment — has primarily focussed on conventional strategies of export-led agriculture (*i.e.,* primary resource specialization) which partly created the debt problem in the first place. In the current international climate the idea of "export your way out of debt" can only help sow the seeds of future famines.

The sub-Saharan African debt problem

Although sub-Saharan African debt does not constitute a threat to western banking system, it is nonetheless a problem which urgently demands a radical solution (Fig. 1). There are no Mexicos or Brazils, no single country that could threaten the international banking system with a default. This indebtedness is crushing all possibilities for economic growth by diverting local resources toward the payment of debt.

In actual terms, sub-Saharan African debt stood at US\$ 161 billion in 1989, an increase of more than 9% over the US\$ 147 billion of the previous year. The ECA estimated the debt for all of Africa to be US\$ 275 billion. However, it is important to note that there are major differences between the debt structures of sub-Saharan Africa and the highly indebted countries, with much higher proportions owed to official bilateral (41%) and multilateral agencies (21%) and much less to commercial banks. While debt to private sources can be rescheduled, debt owed to the International Monetary Fund (IMF) and the World Bank cannot be rescheduled.

By most conventional economic indicators, such as ratio of debt to Gross National Product, sub-Saharan Africa's debt burden was equivalent to 96.9% of its GNP compared to 45.8% for Latin America. For some countries — Congo, Guinea-Bissau, Mauritania, Mozambique and Somalia — the ratio of debt to GNP was 200% or more. In terms of the ratio of debt to exports, the figures were striking: 312% for sub-Sahara Africa as compared to 288% for Latin America. For the sub-continent as a whole, debt per capita in 1989 was US$ 437 as compared to GNP per capita of US$ 449.

Despite a range of debt initiatives by western governments over the past five years, the total stock of sub-Saharan African debt has continued to rise and annual debt servicing costs have eased only marginally. In fact, according to the World Bank, no more than a dozen sub-Saharan African countries have serviced their debts regularly since 1980. Altogether, 25 sub-Saharan countries rescheduled their debts 105 times during 1980–1988.

Up to the end of 1989, a total of US$ 6 billion in official bilateral debt has been cancelled. Given the highly concessional nature of bilateral debt, scheduled debt servicing payment in 1990 were only lowered by an estimated US$ 100 million. Another 17 low income African countries obtained rescheduling from the Paris Club under the more favourable Toranto terms agreed in 1988. The World Bank estimated that more than

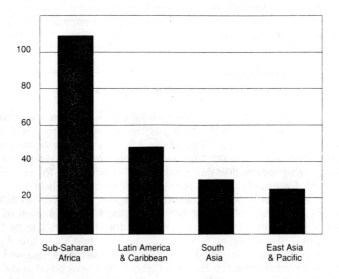

Figure 1. Africa's debt burden (external debt as a % of GNP, 1990). Source: data from World Bank.

US$ 5 billion has been consolidated under Toranto terms since October 1988, resulting in cash flow savings of some US$ 100 million a year.

Such rescheduling may ease immediate cash flow problems, but increases the total stock of debt. About 40% of the long-term non-concessional debt these countries owed to the Paris Club at the end of 1988 represented interest capitalized by Club reschedulings.

I would like to point out that only a small portion of the sub-Saharan African debt qualifies for rescheduling given the debt structure of the continent. Excluded from eligibility are debts owed to the IMF and the World Bank. In fact, sub-Saharan Africa's actual payment to these two institutions accounted for 50.2% of its debt service payments, compared to 17.6% for payments on official bilateral credits and 32.1% on private credits.

The debt problem has reduced the amount of foreign exchange available to purchase necessary imports, leading to a very severe import strangulation, depriving industry, and agriculture of needed inputs, holding back new investment and even the maintenance of the existing capital stock. Debt servicing and the adjustment policies, pushed to release foreign exchange needed to repay the debt, have also worsened social welfare in the areas of health, education, and poverty.

Combined effect of dependence on primary commodities, mainly from agriculture, and the added impetus arising out of being in debt and the need to raise exports, places a heavy strain on the natural resource base of many poor African countries. Governments across the continent are compelled to over-exploit their forests, wetlands, and river basins for short term gain, setting aside considerations of sustainability. In the process, poor farmers lose out as available government resources and personnel are directed to high potential areas where pay-offs are highest.

Structural adjustment: who loses and who gains?

Since the outbreak of the debt crisis in the early 1980s, African nations have become increasingly dependent on World Bank and IMF resources to overcome their financial difficulties. Availability of World Bank and IMF resources is usually conditional upon the implementation of Structural Adjustment Programmes (SAPs), a fact which appears to limit access to these funds and to lower disbursements. This situation is unlikely to change for the rest of the century. It is important to critically evaluate whether stabilization and structural adjustment policies are actually moving African countries towards a desirable economic and ecologically sustainable future.

Almost all the impressive array of research that has come out over the past six years confirms that the orthodox policies being pursued are actually moving African economies away from a desirable medium term and long term path.

While some satisfactory research on the impact of debt/structural adjustment on women and children, *i.e.,* social dimension research, is available, hardly any research with ecological dimension exists — *i.e.,* the impact of debt on sustainable management of natural resources. Here again, the problem has been too much compartmentalization of issues — examining ecological problems as separate activities from rural development strategy.

From the start, we must distinguish between the stabilization policies of the IMF and the structural adjustment policies of the World Bank. Stabilization aims to reduce short term disequilibrium, especially budget deficits, inflation, and B.O.P. deficit. SAPs are concerned with the reorientation of the structure of the economy towards greater efficiency in the medium term (however that is defined!). In reality, however, the distinction between the two sets of programmes has been blurred because World Bank programmes are never instituted unless a fund programme is already in place.

How does adjust-policies work?

When countries cannot pay their debt, the IMF arrives and says, okay, in order to pay back your debt you are going to do all the following things:

Increase your revenues. — Perfectly simple goals, but what do they translate into? Increase your revenues translates into "export even more." You do not have to be an economist to know that this is a "recipe for glut," and for low prices.

The idea is fine if you are the only country in the world told to export more coffee or cotton. But there are now dozens of countries being told to peddle their coffee or sugar in the world market, creating a glut. These countries have a limited range of products that they can sell. When a glut occurs, their whole financial structure collapse. In Ghana, for example, a 50% increase in production of cocoa between 1983 and 1989 was accompanied by a fall in foreign exchange receipts.

The worst part of this approach has been that African governments are cutting down their forests, mining their soils, and devastating their environment in order to earn hard currency.

Reduce government expenditure. — This has been translated into cutbacks in welfare services, on health, education, and transportation. While budgetary balance is important, cuts have been indiscriminate, thereby affecting basic services which are essential for long term development — notably, expenditure on developing human capabilities (health, education, and training), on research and development in priority areas, and on infrastructure, especially in the rural areas. More extension agents and soil conservation experts will not be available in rural areas to assist small farmers unless government investment is accelerated in this area.

Much of the hard work which governments will do to boost agricultural supply in the Sahel through amendments to price and exchange rate policy could be undone easily by their simultaneous cuts in government development expenditure which could dismantle much of the pre-existing structures of support services to farmers.

As described in this brief example, the short-term objectives of structural adjustment are inconsistent with the long-term development needs of African countries. Further more tackling the ecological crisis and the food crisis in the Sahel should not be viewed as separate activities from structural adjustment.

In addition to price and budgetary matters, the impact of other factors on the productive potential of farmers must be examined. For example, in a situation where insecurity of land tenure discourages a peasant from investing the bulk of his/her surplus into agriculture, it is absurd to think that a change of relative price is going to do the trick.

The over-emphasis on market liberalization is based on wrong assumptions and understanding of markets. To begin with, markets have to be created before they can be freed. Markets are not God-given rights, they are the outcome of social struggles. There will not be a place for small farmers in a free market system unless they increase their leverage to exert influence over decisions. Only then can markets be structured to ensure the allocation of resources to small farmers — from credit to foreign exchange. For example, a free market economy will not be up and running tomorrow or the day after just because the president signs a legislation! Different social coalitions will reassess their relative positions in the society to either take advantage of policies or to sabotage them.

Immediate impact of debt on sustainable agriculture

The debt crisis and the management of the crisis through structural adjustment policies have the following consequences:

Reinforcing dependence on primary products. — Despite a major turbulence in the world economy, the prescriptions offered by the MDBs to readjust African economies are based on the positive scenario that 1) world trade will grow, 2) protectionist measures will not be imposed and 3) that commodity prices will go up. The reality has been quite different. Commodity prices are now at their lowest in 50 years, making it ever harder for African countries to "export their way out of debt." Trade barriers intensify the problem. These policies are an extension of the historical process and reinforce the position of African countries as "dessert and fruit-cocktail" economies. Further, the policies being advocated ignore African environmental reality. By pushing mono-cropping instead of inter-croping, fostering tractor ploughing on soils that should have been disturbed as little as possible, the sustainability of export oriented agriculture projects has always been in jeopardy owing to problems of declining soil fertility, sedimentation and salination, and water supply problems.

Increased vulnerability of African economies. — Commodity decline, shifts in world consumption patterns and technical innovation in synthetic substitutes, food chemistry, and bio-technology have cut demand for traditional raw materials — bad news for the countries dependent on one or two commodities. Structural adjustments are virtually impossible to attain within an environment of uncertain commodity situation.

Switching effect. — These reform measures are contributing to deforestation and soil erosion by preempting productive resources, money, and valuable personnel, research and development institutions, technology, and credit toward the export producing sector for short term gains. Meanwhile, small farmers go on exploiting degraded land for lack of technical support in agro-forestry, soil conservation, and food production.

Perpetuation of rural poverty and inequality. — It is a well known fact that grassroot organizations in the Sahel region have taken a leading role against environmental degradation. Using the traditional concept of selfhelp, villagers have been doing conservation work — from Burkina Faso to Kenya. Unfortunately, these dynamic groups are rarely the subject of conventional adjustment programmesmes. Government support to agriculture has mainly favoured large farmers who produce for export to the disadvantage of small farmers who produce food for local consumption. This perpetuates rural poverty and inequality.

In Ghana, for example, the agricultural sector reform has primarily focussed on cocoa production since the early 1980s. A closer examination of the distributional effect from cocoa earnings reveals a different story. Cocoa farmers comprise only 18% of Ghana's farming population and are concentrated primarily in the south. A 1987 Overseas Development Institute, University of Ghana, study revealed that 32% of the cocoa farmers in the Ashanti region received 94% of the gross cocoa income, while 68% of the farmers received only 6%.

Meanwhile, the per capita income of non-cocoa farmers has stagnated. With the exception of 1984, Ghana's rate of food self-sufficiency has been steadily declining because incentives are not available to food producers. Scarce resources such as credit, extension, technology, and other necessary inputs have been preempted to the export sector while poor peasants are left to fend for themselves. Further, drastic devaluation of the Cedi has made the cost of inputs such as fertilizers more expensive for ordinary peasants to afford.

Weakening skill levels. — Adjustment programmess ignore the managerial and economic realities of African governments. Implementation of structural adjustment programmes has progressively weakened the skills base through a withdrawal of resources from education and through an acceleration of the brain drain from the continent. A joint ECA/ International Labour Organization report estimated that 70,000 mid-level and high-level Africans left the continent in 1987, up from 40,000 in 1985. This represents approximately 30% of Africa's skilled human resources. The decline in real earnings, the removal of subsidies, the increase in the cost of medical, educational and other services, the freezing of wages, and the massive retrenchments from the public service have made life difficult for all Africans. With very few options left, many skilled Africans are forced to leave the continent while the poor peasants go into the underground economy and illegal cross-border trading.

Another orthodox irony is that the donors attempt to use the state to implement orthodox policies while trying to get the state to dismantle itself!

Intensification of resource use. — In countries that are dependent on one or two primary products, diversification is essential. Adjustment policies, however, have tended to push the wrong kind of diversification strategies, by forcing governments to exploit their forests. A case in point is Ghana. To make up for declining foreign exchange earnings from cocoa, the timber industry is being revived with support from the World Bank. Timber output rose from 147,000 cubic meters to 413,000 cubic meters between 1984 and 1987. This has accelerated the steady destruction of Ghana's forests whose size have decreased considerably as a result of decades of agricultural conversion of forest land. Environmentalists predict that if timber production is maintained at its current rate without appropriate environmental controls and if uncontrolled tree felling continues, the Ghanaian countryside could be stripped naked by the year 2000. As it has happened in other parts of Africa, this would lead to a disastrous situation of reduced food production, declining soil fertility, and water supply problems.

What is to be done?

The greatest challenge for sub-Saharan Africa is how to transform crushing social needs into effective demand and then to meet that demand by turning first to domestically produced goods and services, next to the region, and only after that to the wider world.

The ecological crisis in sub-Sahara Africa cannot be understood in isolation from the structural processes that are determining the way natural resources are used. While such policy reforms as devaluation, pricing policy, and budget and tax reforms are necessary components of a balanced and integrated national development strategy, they alone will have little impact on sustainable agricultural production, sound management of natural resources, and in reducing poverty and inequality unless accompanied by fundamental transformation of economic and political structures. Although the task of transformation could take a long time before it is achieved, the following suggestions should be adopted in the meantime to lay the foundations for long term sustainable development in sub-Saharan Africa.

Land reform and rural development strategy. — Sub-Saharan African governments and donors must give priority to the improvement and diversification of agricultural activities and the elimination of poverty and inequality through the establishment of sustainable production patterns geared to improve the living conditions of the poor. Special emphasis should be placed on food production for local consumption by changing the incentive structure in favour of small farmers. Similar measures must be taken to strengthen the capacity of farmers to undertake environmental protection. To achieve these goals, the political will must be found to reorganize the agrarian structure so that land is distributed equitably and the security of tenure clarified through legislation.

Institution building. — Agricultural supportive services, such as extension, research and agricultural marketing, and credit are ineffective in most sub-Saharan African countries. These services must be strengthened both in the allocation of resources and in the provision of facilities and equipment to strengthen their analytical and service delivery capability. This can be achieved through institutional support in the training of extension agents, sectoral and management planning departments and in improving logistical support.

Access to infrastructure and services. — As shown in this paper, poor farmers are often isolated and lack access to basic services that could improve their productivity. Lack of access to credit, extension service, fertilizers, storage, roads, and other services has been the major impediments to increased agricultural production and improved management of resources. Improving their access to these services will go a long way to arrest the process of land degradation and declining agricultural output in Africa.

Supporting local NGOs. — Government agencies cannot do all what is expected of them. Special efforts must be made to upgrade and employ services of local non-governmental organizations, cooperatives and women's organizations, better placed than public agencies, to stimulate a process of change. But for these local organizations to succeed, both national governments and donors must make supporting services, from training to technical assistance, available to them all. Training of peasants in various techniques of agro-forestry and soil and water conservation must be expanded. This requires setting up new demonstration centres close by as well as the provision of seedlings and other inputs.

Agricultural research for small farmers. — Research should be reorientated toward solving the problems facing peasant agriculture by involving the peasants themselves in decision making and by tapping their knowledge of their natural environment. Policy-makers need to expand the scope of their understanding of the various aspects of peasant agriculture. Research in drought resistant crops or high yield-increasing agricultural innovations can reduce the pressure on marginal lands. Improving the traditional tilling techniques and adjusting them to soil conditions, and the introduction of new and improved methods of soil and water conservation should be accorded top priority.

Addressing the problem of women farmers. — Women are responsible for 60–90% of food production, processing, and marketing in African countries. Women's organizations have taken a leading role against soil erosion and deforestation. Yet women have the least access to improved technology, credit, extension services, and land. Thus, programmess to increase food security and preserve the natural resource base in Africa must reach women if they are to succeed. Efforts must be made to improve women's access to productive resources, and by reorientating agricultural training as well as other supportive services to solve the problems of women farmers.

Easing fuelwood shortages. — Efforts must be made to introduce efficient stoves suitable for specific locations, supported by strong extension, demonstration, and promotional activities. The technology introduced must be simple and easy to reproduce using local material, and must be affordable for poor people to purchase one. This will drastically reduce the use of agricultural residue for cooking purposes.

Family planning programmes. — Population growth and the shortage of arable land have compounded the environmental crisis in Africa. The lack of family planning services to the poor is one of the main reasons for high pregnancy. Both donors and national government should make available increasing resources to strengthen existing population activities and family planning services while trying to improve access to productive employment. Raising the education status of women and girls should also be accorded top priority.

Diversification of economic activity. — Beside increasing opportunities in small-scale agriculture, attention must be paid to generating and expanding employment in non-agricultural sectors. Required action include the intensification of public works, such as secondary roads, reforestation, soil conservation, clean water supplies, rural electrification, health clinics, schools, and agro-service centres. These activities would strengthen the internal working of the national economy by stimulating production and consumption of local goods and services.

Easing the debt burden. — Since 1986, creditor nations have implemented different measures to ease the debt burden of sub-Saharan Africa. Unfortunately, the measures taken up until now have been inadequate and require further expansion. Consistent with the recommendations of the United Nations Secretary-General, cancellation of all official bilateral debt and other semi-official debts, such as export credits, should be considered. Since most of sub-Saharan African debt is owed to multilateral institutions, ways should be found to substantially retire much of this debt. Debt relief, however, must be conditioned on political democratization and on the reorientation of national resources toward ecologically and socially sustainable forms of agricultural production, and greater involvement of the poor in economic decisions and development activities affecting their lives.

Fostering government accountability. — In the final analysis, however, solutions to the ecological crisis in sub-Saharan Africa must address social and political issues as well as economic and technical ones. It is not possible to eradicate poverty and inequality, and preserve the natural resource base without transferring some real power to the marginalized people. The deepening of democracy and the empowerment of people at the local level are a *sin quo non* to sustainable development. Without a democratic political environment, all the measures adopted in the interim by donors, local governments, NGOs, or community groups will be inadequate to resolve Africa's economic and ecological crisis.

Development of Indigenous Integrated Farming Systems in the Sahel

Mike Speirs
Copenhagen, Denmark

"Les sociétés sahéliennes (...) moins favorisées par le ciel (que des pays comme la Côte d'Ivoire) n'ont bénéficié que de rentes modestes et très provisoires, rentes qui ont été surtout utilisées à l'achat des biens de consommation et fort peu à investir pour rendre l'appareil productif plus efficace (encore moins que dans le cas ivoirien). L'aide extérieure a joué pour elles un rôle assez analogue à celui de la rente, utilisée de plus en plus pour consommer et de moins en moins pour investir. L'ajustement structurel a sans doute aussi assaini les économies sans créer le choc qui aurait déclenché le développement. Il est surtout ressenti par les élites dirigeantes comme une tracasserie supplémentaire que les Occidentaux (gens biens compliqués!) mettent au maintien de leur aide. Et il est ressenti aussi comme une menace pour leur position privilégiée. La réponse des sociétés sahéliennes est plutôt de s'enforcer dans une économie informelle peu productive et dans un système traditionnel de dons et de contre-dons qui, la dureté des temps aidant, a pris une nouvelle vigueur." (Giri 1991).

Introduction

The extensive debate on rural development in Africa involves policy makers, development workers, researchers and donor agencies in various ways. The aim of this paper is to outline some of the economic policy issues which are faced by governments and decision makers in the Sahel in efforts to promote rural development based on sustainable farming systems. Although, as Hirsch (1990) argues, there may be a considerable gap between agricultural policy pronouncements by governments and the actual implementation of price, tariff, taxation and exchange rate reforms, it is useful to examine the broad contours of economic development strategies in this region of West Africa. Three inter-related development objectives are implicit in the discussion which follows.
    Firstly, increasing food security is a major policy concern in the Sahel, where many people suffer from hunger, malnutrition, and, in bad years such as 1973–1974 and 1984–1985, from widespread starvation.

While Kennes (1990) notes that food security can be defined simply as the "absence of hunger and malnutrition," a more thorough description of the concept includes the idea of "access by all people at all times to enough food for an active healthy life" (World Bank 1986), which in turn suggests both that adequate supplies are available and that people have the means to acquire food. Economic analyses of food security further emphasise price factors (agricultural production costs and consumer food prices), while nutritional concerns are stressed in terms of food quality, and the ecological approach to food security is concerned with sustainable agricultural production techniques. Some investigations of food security focus on access to food by households in specific locations, others deal with national or regional food security. Implicit in all the various definitions of the concept is a concern with the living standards and welfare of the poor, and with measures which may contribute to improving both supplies of, and access to, healthy and nutritious food.

The provision of food aid is one simple means of raising food security levels in deficit areas, but there are many pitfalls in the distribution process, and in non-emergencies other forms of development assistance, linked to agricultural policy reforms, may be more appropriate. As the data presented indicate (Tab. 1), many Sahelian states have received large quantities of food aid in addition to cereals imports (mainly of wheat and rice) during the past decades. Declining imports and a considerable reduction in the flow of food aid at the end of the 1980s were largely attributed to the good harvests in the region in this period, though contracting demand under adjustment programmes and the

Year	Production	Imports	Food aid	Imports and food aid as % of total cereals available
1960	4,300	2,000	—	4
1970	5,200	540	—	9
1980	5,500	919	—	14
1981	5,821	1,088	—	16
1982	5,383	1,207	521	24
1983	4,861	1,393	426	27
1984	4,253	1,792	629	36
1985	7,383	1,794	1,142	28
1986	7,593	1,279	563	20
1987	6,722	931	366	16
1988	8,867	966	322	13
1989	8,011	950	263	13
1990	7,543	—	—	—

Table 1. Estimates of cereals production, imports and food aid in the Sahelian (CILSS) countries (1000 tonnes). Sources: CILSS and FAO data. Note: these estimates include food aid shipments.

lack of foreign exchange to pay for imports may also have been contributing factors.

It has been argued that the food security of the populations in the region can best be ensured through reliance on such imports while efforts are made to develop the "competitiveness" of the agricultural sector using other policy instruments. Increasing self-sufficiency in the production of food grains is another approach to improving food security levels. Net food importing countries, however, are faced with a dilemma in the sense that the costs of positive protection of the agricultural sector (to raise domestic supplies) are counterbalanced by the possibilities of obtaining foreign exchange through exports (of crops and manufactured goods) and the risks of international supply shortages (of food grains). The links between food security, marketing and international trade in agricultural products are of crucial importance to development prospects in the Sahel.

Clearly, and this is the second objective, increasing incomes in rural communities through both agricultural and non-agricultural activities are a vital part of the development process in the Sahel and elsewhere in Africa. One of the explicit aims of "structural adjustment" programmes has been to revitalise the agricultural sector by focusing on elements of "economic mismanagement" by governments and para-statal organisations in Africa which, it is argued, led to falling food and cash crop output and deteriorating living standards in rural areas. Price and exchange rate changes, reduced state intervention in crop marketing and "incentives" designed to encourage private sector entrepreneurs have formed the cornerstones of these programmes.

The design and implementation of economic policies which tackle the underlying causes of stagnating agricultural production imply a broader approach. The diversity and fragility of natural resources, dispersed settlement patterns and extensive farming systems, insecure access to land, and the low level of technological development have been identified amongst the many constraining factors or "structural deficiencies" which result in low levels of agricultural productivity and output in Africa (Platteau 1988). Improving levels of food security, income and employment in rural areas necessitate an emphasis on enhanced productivity in the smallholder (peasant) sector, backed up with investment in better production techniques, infrastructure and agricultural support services. Thus, for example, Streeten (1987) has identified six essential "ins" of agricultural development: incentives (prices), inputs, innovation (technology), information, infrastructure and institutions. "Getting agricultural prices right" is a necessary, but not sufficient, condition for agricultural development.

A third development objective, ensuring effective land use management through the development of integrated farming systems (such as "agro-sylvo-pastoral" schemes) has become an important priority, given the increasing severity of environmental degradation in the Sahel. Traditional extensive farming systems which reflected the adaptation to low and variable rainfall, poor and fragile soils and the relative abundance of land, have come under increasing pressure. The total cereals production increases in the four Sahelian countries, Senegal, Mali, Burkina Faso and Niger, are largely due to an increase in the cultivated area rather than to yield gains per hectare (which only slightly increased during the 1980s). More and more pasture has been put under cultivation and fallow periods have shortened.

Traditional semi-nomadic and transhuman animal production systems have been undermined in this process, as access to grazing land has become more difficult and frequent conflicts between pastoralists and crop cultivators are observed. Although there is evidence that many arable farmers make use of animals for traction and cultivation purposes, it does not appear that the adoption of integrated farming systems (which entail spreading manure on the poor soils in the semi-arid areas) can prevent further declines in soil fertility. Indeed, an extension of the cultivated area is often encouraged by the use of draught animals in cereals and cash crop production. Adopting and maintaining mixed farming systems appear to be a viable prospect principally in cash cropping areas, notably in the cotton zone of south-western Burkina Faso and southern Mali.

The development of sustainable agriculture to reverse the various forms of land degradation prevalent in the region (over-grazing, declining fallow periods, erosion and deforestation) depends on financial and human investment in improved production techniques, which in turn requires security of access through land tenure reforms, and a favourable "economic climate" in which agriculture is a profitable activity. The introduction of land use management schemes and reforms of tenure legislation have recently been included in the structural adjustment programmes agreed between the governments of the Sahel and major aid donors. But the adjustment process itself, originally conceived as a set of short-term measures to reduce balance of payments and budgetary deficits, appears to have a long way to go.

Comparative disadvantages

Although a transformation of traditional agriculture in the Sahel is underway, crop yields, output and labour productivity remain low. At the same time, the increasing dependence on staple food imports represents a long-term threat to the development of indigenous agriculture in the region. Furthermore, the terms of trade for key West African export commodities including coffee, cocoa, cotton and groundnuts have deteriorated throughout the 1980s, and foreign exchange earnings have declined. Yet despite these difficulties, over 75% of the Sahelian population continues to depend on agriculture and, apart from the emergence of the informal sector, small-scale manufacturing and commerce there are few signs of an industrialisation process which could absorb the expanding labour force. Rapidly and simultaneously increasing economic growth rates on the basis of agricultural output, satisfying the food security needs of a growing population and transforming the systems of production themselves, are an awesome task.

In a sense the problem is that, given increasing competition on export markets, the low levels of productivity and the high costs of intensifying agricultural production in ecologically fragile zones such as the Sahel, together with the weakness of infrastructure and institutions, many countries appear to face a complex set of comparative "disadvantages." Thus the extent to which the reforms which have been introduced through structural adjustment programmes can be expected to "deliver the goods" in terms of higher growth rates and standards of living, as well as the relationship between these liberalisation attempts and the forces of political change are central issues. In addition, the roles of external agencies in agricultural development and the pattern of insertion of African economies in the world trade system further determine development patterns and the range of opportunities in policy changes. The following brief descriptions of the dynamics of two product markets in West Africa illustrate the complexity of the economic policy issues which arise in the development of sustainable farming systems.

Table 2 shows estimates of changing self-sufficiency ratios for several key agricultural commodities in West Africa. Almost all coarse grains consumed in the region (millet, sorghum and maize) are also produced locally, whereas 40% of the rice consumed and 96% of the wheat were imported (in the period 1985-1987). The scale of the potential, but declining, beef surplus (excluding Nigeria) is also shown in the table.

In an assessment of cereals marketing and trade, Berg (1989) notes that the development of low cost production systems in other regions of the world (such as North America and Asia), together with subsidised agricultural exports by major producers (including the EEC), and "exchange rates and other policies in the Sahel which work to the disadvantage of domestic producers and to the advantage of imports" present serious challenges to the competitiveness of agricultural production and to the improvement of food security in this region. Since the end of the 1970s, the domestic prices of "traditional" cereals in the

Year	1979–81	1985–87	1979–81	1985–87	1979–81	1985–87
Wheat	1	4	0.5	2	2	7
Rice	59	60	59	54	58	73
Coarse grain	98	99	98	98	97	99
Beef	99	100	136	113	65	83

Table 2. Self-sufficiency ratios in West Africa.

Sahelian countries, which are rarely traded on world markets, have risen relative to imported grain. Rice prices on world markets have fallen in the same period, relative to both coarse grain and to manufactured goods (Delgado 1989a,b). Consequently, it has been argued that a protected cereals market should be established in West Africa, to encourage the development of indigenous farming systems based on "traditional cereals" as an alternative to increasing dependence on food imports.

But the results of a number of household consumption surveys in several Sahelian countries indicate that higher grain prices due to the imposition of tariffs (or other protective measures) aimed at improving the competitive position of domestic cereals production by reducing imports, are likely to have a negative impact on consumers in both deficit rural areas and amongst the urban poor. Increases in rice and wheat consumption, especially in the urban areas of both the Sahelian and the coastal countries, do not appear to be driven by price factors, but by the process of urbanisation itself, notably in terms of the lower processing and preparation costs of rice. In the rural areas of the Sahel, an increasingly large proportion of households are no longer simply subsistence producers in the sense that net cereals purchases have become an important source of food supplies. Coarse grains are grown to feed the household, not to generate cash income.

Stable prices and guaranteed marketing outlets for cash crops like cotton are important factors which influence smallholder production decisions. Nevertheless, in the rural areas where monetary incomes are restricted, the relative prices of different cereals may directly influence consumption patterns. It has also been pointed out that higher cereal prices will have a significant impact on production costs in all sectors, by increasing the price of the most important wage good, *i.e.,* food. This is the classic food price dilemma (Streeten 1987). Promoting growth in the agricultural sector through price policies alone may hinder economic growth in other sectors by driving up the cost of labour.

Further investigations of the comparative advantages of different production options suggest that, given the high costs of investment in irrigation and the difficulties of transportation, the production of cotton as well as rainfed cereals and livestock are likely to continue to offer opportunities for expansion in specific zones. Reducing the production

Year	1970	1980	1987
Exports (from Burkina Faso, Mali, Niger, Chad and Central African Rep.)	646	670	457
Imports (Benin, Ivroy Coast, Ghana, Liberia, Nigeria, Sierra Leone and Togo)			
— live animals from the Sahel	700	689	478
— meat from outside Africa	124	370	740

Table 3. Meat and cattle trade in West Africa (1000 head). Source: Josserand (1990).

Year	Mali	Abidjan (live cattle)	Abidjan (cif Argentina)	Abidjan (cif EEC *capa*)
1980	485	580		—
1981	550	650	412	475
1982	570	680	393	400
1983	530	700	437	300
1984	550	640	528	250
1985	500	775	507	250
1986	600	740	460	180
1987	660	750	465	220
1988	640	700	400	280
1989	620	740	—	—

Table 4. Meat prices in West Africa (annual average. F.CFA./kg). Source: Josserand (1990).

and transaction, *i.e.,* trading and marketing, costs in these sectors could enhance the possibilities for exports to the coastal countries, although there appears to be a limited demand for millet and sorghum in the urban areas of the forest zone. On the other hand, investment in maize production, both in the southern Sahel (Sudano-Guinean zone) and in the coastal countries, is likely to yield positive results, given the attractiveness of this product to consumers as well as the successful implementation of intensification schemes and productivity increases in various locations (Dione 1990, Reardon 1989).

Similarly, there is an urgent need to recapture coastal markets for Sahelian livestock, given the prospects for expanding both cattle and small ruminant production in the northern zones of the CILSS countries. As shown in Table 3, meat imports accounted for an increasing share of the West and Central African market from 1970 to 1987.

Low effective demand for livestock on national and regional export markets, and fierce competition from non-African (and highly subsidised) meat imports are progressively reducing the regions meat exports despite the high production capacity in the Sahel at costs comparable (exclusive of subsidies) to those of other producers. Table 4 shows some price data for selected markets, indicating the significant price differential in the 1980s between live cattle imported by a coastal country (Ivory Coast), and frozen meat (*capa*) from the EEC. Competitiveness of Argentinian beef exports is also illustrated in the table.

Protection of the Sahelian meat producers, and the ending of dumping on West African markets, are two important measures that could encourage regional production, as recent studies have suggested[1]. The reduction of transaction and marketing costs through improved

---

1 See, *inter alia,* the studies by Delgado (1989c), Josserand (1990) and Kulibaba and Holtzman (1990), the latter focusing on meat and animal product trade between Mali, Burkina Faso and Ivory Coast.

transport and the relaxation of taxation and import controls by the coastal countries are additional elements of policy reform, such that the comparative advantages enjoyed by the livestock sector in the Sahel can be fully exploited. As in the case of the regional cereals market, however meat consumers would face higher prices in a protected market.

Markets, states and trade

As the consequences of neglecting rural development became apparent through detailed investigations of the constraints which affected the agricultural sector in many African countries during the 1970s and early 1980s, the idea has gained ground that less state intervention means better agricultural performance as well as better government. However, further analyses of food security in terms of ensuring an adequate supply of food as well as access (entitlements) through income generation, suggest that market solutions based on increases in producer prices and restrictions on para-statal intervention are not sufficient responses to the agrarian crisis. Given the predominance of imperfect markets, insecure property rights, and the widespread use of political influence in allocating resources, as well as the emergence of extensive parallel marketing networks, privatisation and liberalisation through structural adjustment programmes may do little to improve the bargaining position and food security of the poorest farmers whose needs are not backed by effective demand.

A number of studies have concluded that an effective agricultural marketing reform is closely linked to the abilities of both private sector agents and state institutions to redefine their roles and carry out complementary functions (Commander 1989, Lele and Christiansen 1989). In Burkina Faso for example, while it is increasingly recognised that private merchants are efficient and competitive in many respects, strengthening cooperative forms of exchange (through the "Office National des Cereales," and the network of village level cereals banks) can also contribute to ensuring that peasant farmers are able to maximise their incomes through the sale of cereals and simultaneously maintain consumption levels during the *soudure* (hungry season). As Killick (1989) noted, "state-private relationships are generally symbiotic.[2]"

Although changing the objectives and methods of state intervention in domestic food and agricultural markets in the Sahel may lead to reductions in the budgetary deficits incurred by the para-statals and to improvements in efficiency, the long term problems posed by the declining competitiveness of the agricultural sector, at the interface

---

[2] "If liberalisation is to lead to improved market performance, it is not enough that the state stop doing certain things, like running monopoly marketing systems. The state must also take on new responsibilities, such as providing public information systems and improving credit markets, in order to facilitate the private sector's ability to respond to its new opportunities" (Staatz *et al.* 1989).

between West African markets and the world economy, must also be faced. Shifting from extensive to intensive farming practices in order to increase the productivity of labour and land, and encouraging the progressive commercial integration and modernisation of agricultural production, are elements of this process (Platteau 1988). Similarly, other domestic policy reforms which tackle the difficult problems of land tenure, access to productive resources, *i.e.,* inputs, water supplies, pasture, forests, and so on, as well the efficient management of these production factors, are important measures in terms of both productivity and equitable income generation.

But these approaches may necessitate protection against the import of competing food products, such that it is advantageous and profitable to invest in increasing the production of traditional cereals which can be sold to an urban population whose consumption patterns are rapidly changing. The risk is not only that the costs of protecting domestic markets may be very high, particularly in terms of rice supply, but also that many poor food consumers, including rural households in areas with production deficits due to environmental degradation, may suffer from such an expensive food strategy. Nevertheless, the substitution of domestically produced cereals by imported rice and wheat has become one of the major issues which must be tackled through trade policy reform in West Africa. Without some forms of protection, preferably involving the coordination of agricultural policies at the regional level, it is difficult to envisage productivity and output increases in the Sahel.

Left open to international competition, the agricultural sector will continue to be undermined by cheap food imports, while the prospects for improving efficiency of export crop production will remain uncertain. Expanding the production and export of specific agricultural commodities will thus entail thorough assessments of the changing pattern of demand in different parts of West Africa in order to locate potential markets, as well as stepping up efforts to resolve serious environmental difficulties which constrain supply increases. Coherent regional agricultural policy, entailing the harmonisation of national development programmes and economic policies, including tariffs, investment schemes, infrastructure and marketing, could contribute to maximising the benefits of production complementarities in the different agro-ecological zones of West Africa[3].

At the same time, the shift towards a more liberal agricultural trade regime in the course of the Uruguay Round of international GATT negotiations has a potentially significant impact on development

---

3 The pros and cons of cereals market protection at national and regional levels continue to be investigated and debated (Delgado 1989a,c, Gentil 1989). Similarly, the difficulties of regional cooperation in West Africa, and market integration through the harmonisation of economic and agricultural policies have been thoroughly assessed (Bach and Vallee 1990, Berg 1991, Egg *et al.* 1991). Meanwhile agricultural ministers from 15 countries met in Dakar in March 1991 to draw up a new initiative focusing on the development of regional trade in cereals, oilseeds and animal products and on the improvement of export crop production systems.

programmes and policies in West Africa and elsewhere. The latest round of multilateral negotiations within the framework of the General Agreement on Tariffs and Trade were initiated in Uruguay in 1986. The major agricultural super-powers, including the EEC and the USA as well as the Cairns Group of new food and agricultural exporting nations remained deadlocked on the issue of reducing price support and farm subsidy levels throughout the four years foreseen for concluding an agreement.[4] Needless to say, farming communities in Europe, Japan and many other parts of the world have mounted considerable opposition to proposals to reduce support levels and to abandon guarantied price systems.

As far as West Africa is concerned, and particularly those countries which have become net food importers, the GATT articles dealing with special and differential treatment for underdeveloped countries are of considerable importance. Without these provisions in the General Agreement, any form of trade distorting barrier (such as import quotas, tariffs and price subsidies) can be outlawed by the contracting parties, since these measures are contrary to the operation of liberal, free-market principles in agricultural trade. But it is the practice of subsidising exports on world markets, at prices below the domestic costs of production in the EEC, the USA and elsewhere, which is particularly harmful to producers of commodities such as traditional cereals, milk and meat in regions like West Africa.[5]

Agricultural policy changes in West Africa will be directly affected by the outcome of the GATT negotiations in the Uruguay Round. It is essential to maintain the provisions for special and differential treatment of developing countries in the final agreement, such that net food importing countries in West Africa are able to enhance food security using protective measures which act as incentives to increase production and improve welfare.[6] It is also necessary to reach an agreement to

---

4   The Cairns groups includes Argentina, Australia, Brazil, Canada, Indonesia, Malaysia, Thailand and New Zealand. For an introduction to the GATT and the Uruguay Round and for an assessment of the strategies adopted by the "contracting parties" during the negotiations, see the paper by Watkins (1991).

5 Numerous attempts have been made to estimate the effects of agricultural trade liberalisation in the OECD on different groups of countries. The papers edited by Goldin and Knudsen (1990) are particularly interesting. In addition to the political difficulties associated with trade liberalisation, it is argued that poverty and income distribution within African countries are affected by many more significant factors than simply the gains and losses which might result from reduced levels of protection in the OECD countries (Raikes 1988).

6 Watkins (1991) notes that the Jamaican government has called for "GATT rules which recognise a fundamental distinction between subsidies in the north (which finance over-production and cause world market distortions, which should be subject to policy disciplines) and subsidies designed to increase food self-reliance, protect rural employment and promote ecologically sustainable farming (which are properly matters of national policy sovereignty)." But these demands may prove "too little too late."

phase out export subsidies which will improve the competitiveness of livestock production in the Sahel. Reducing price support in the EEC will probably lead to higher world cereals prices, which will be a drawback in that cereals imports by West African countries will become more expensive. But compensatory measures (including targeted food aid) could be introduced in a transitional period, as producers respond to better marketing conditions. The issue of improving market access in the OECD for export products from West Africa is also important. In these ways, a trade agreement which includes provisions for food security and environmental protection could be beneficial to agricultural producers in a marginal area like the Sahel.

Some concluding remarks

Encouraging the development of indigenous farming systems in the Sahel is an essential step in economic policy reform designed to achieve sustainable growth in this part of Africa. But the political difficulties inherent in the process of shifting the balance of power towards smallholder (peasant) producers cannot be underestimated. It may be that the present wave of democratisation in West Africa will enable greater empowerment of the rural populations, who have often suffered from marginalisation and the negative effects of "top down" development strategies. But the full implications of the present trend towards trade liberalisation and privatisation through adjustment programmes, which affect marketing arrangements, agricultural sector support services and land use management initiatives have yet to be fully understood (Giri 1991). Nevertheless, a political "space" has opened up in which peasant farmers, pastoral associations and those concerned with appropriate and sustainable forms of rural development may be able to exert greater influence.[7]

On the other hand, as Raikes (1988) argued, the peasantry as such is unlikely to form the basis for a major change in political structure. One of the dangers of adjustment programmes which undermine state institutions in politically fragmentary nations such as those in West Africa, is that the weaker and marginalised producers (including overworked women farmers in remote rural areas, pastoralists, nomads and other groups) who have had few opportunities to organise and formulate specific demands for improvements and policy reforms, will remain at a distance from the centres of power and decision making.

---

7 Half-hidden in the title of this paper is an acknowledgement of the remarkable study by Richards (1985) who presents a strong case for supporting indigenous initiatives in West African agriculture, arguing that: "streamlining agricultural support services in order to cope with large current account deficits (in addition to the acute external debt problem) ... represents something of an opportunity to experiment with incremental approaches to agricultural Research and Development and the peoples science option."

Thus, for example, in a discussion of privatisation and land tenure legislation in the Sahel, Faye (1990) points out that although private ownership may increase incentives to invest in conservation and protection of the environment, the subsequent competition for control over natural resources entails a risk of greater inequality and a deterioration in the conditions of the poorest populations (who are unable to gain access to land, *etc.*).[8]

Although centralised and authoritarian systems of resource allocation built on patronage and kinship relations around state institutions may be on the decline, the search for "farmer first" agricultural development strategies in the complex, diverse and risk prone semi-arid zone is fraught with difficulties.[9]

A significant contribution can be made by further research into indigenous farming systems, and by drawing attention to the distortions of economic policy linked to inequalities of access and influence. Similarly, the development of applied research programmes which aims to assess the dimensions of food security problems in the Sahel and formulate appropriate policy reforms is an important step forward (Dione 1990, Maxwell 1990). But the key role to be played by external agencies in the development of sustainable agriculture in the Sahel is to combine support for organisations and producer associations with a consistent policy framework (including trade agreements) which will enable the farming populations of the region to overcome comparative disadvantages and increase standards of living.

Literature cited

Bach, D. and Vallée, O. 1990. L'intégration régionale — espaces politiques et marchés parallèles. — Politique Africaine, 39.

Berg, E. 1989. The competitiveness of Sahelian agriculture. — In Club du Sahel, CILSS (eds.), Regional cereals markets in West Africa — a compilation of studies on Sahelian agriculture, regional trade and world markets. D(89)337, Club du Sahel, CILSS, OECD, Paris.

Berg, E.1991. Strategies for West African economic integration — issues and

---

8 See also, *inter alia*, Monimart (1989) who considers the difficulties faced by "les femmes desertees" in the Sahel, and the survey of farmers' organisations by Snrech (1988).

9 "The essence of farmer first is that it reverses some parts of the transfer of technology process which have tended to go unquestioned. A reversal of explanation looks for reasons why farmers do not adopt new technology in deficiencies in the technology and the process which generated it, rather than in farmer's ignorance. Location and roles are reversed, with farms and farmers in control instead of research stations, laboratories and scientists. In Africa it has usually been the non-governmental organisations (NGOs) which have carried out the most innovative work in this field. Limited resources within the national research and extension services and the hierarchical structure of the latter have both constrained the extent to which staff have been able or willing to maintain much contact with farmers, especially those in the more distant, marginal areas" (Toulmin and Chambers 1990).

approaches. — D(91)382, Club du Sahel, CILSS, OECD, Paris.

Commander, S. 1989. Structural adjustment and agriculture — theory and practice in Africa and Latin America. — ODI, Currey, London.

Delgado, C. 1989a. Questions à propos d'un espace régional protégé pour les céréales au Sahel. — Economie Rurale, 190.

Delgado, C. 1989b. Why is rice and wheat consumption increasing in West Africa? — Unpublished paper presented at a seminar of the European Association of Agricultural Economists, Montpellier.

Delgado, C. 1989c. Cereals protection within the broader regional context of agricultural trade problems affecting the Sahel. — Unpublished paper presented at the Club du Sahel, CILSS seminar on regional cereals markets in West Africa, Lomé.

Dione, J. 1990. Sécurité alimentaire au Sahel — point sur les études et projet d'agenda de recherche. — Document de recherche 90-02, PRISAS, Institut du Sahel, Bamako.

Egg, J., Igué, J. and Coste, J. 1991. Approaches to regional cooperation in West Africa — discussion based on work conducted by INRA/IRAM/UNB. — D(91)387, Club du Sahel, CILSS, OECD, Paris.

Faye, J. 1990. Le control privé permet-il une meilleure gestion des ressources naturelles? — Document non-publié présenté au séminaire "l'avenir de l'agriculture des pays du Sahel", Club du Sahel, CIRAD, Paris and Montpellier.

Gentil, D. 1989. Production agricole, échanges régionaux et importations au Sahel. — Document non-publié présenté au séminaire du Club du Sahel, CILSS sur les espaces céréaliers régionaux en Afrique de l'Ouest, Lomé.

Giri, J. 1991. Les années 1980 dans le Sahel — un essai de bilan. — D(91)381, Club du Sahel, CILSS, OCDE, Paris.

Goldin, I. and Knudsen, O. 1990. Agricultural trade liberalisation — implications for developing countries. — OECD, Paris.

Hirsch, R. 1990. Adjustment structurel et politiques alimentaires en Afrique sub-saharienne. — Politique Africaine, 37.

Igué, O. 1983. L'officiel, le parallele et le clandestin — commerces et integration en Afrique de l'Ouest. — Politique Africaine, 9.

Josserand, H. 1990. West African systems of production and trade in livestock products — Issues paper. D(90)351, Club du Sahel, CILSS, OECD, Paris.

Kennes, W. 1990. The European Community and food security. — IDS Bulletin, 21-3.

Killick, T. 1989. A reaction too far — economic theory and the role of the state in developing countries. — Overseas Development Institute, London.

Kulibaba, N. and Holtzman, J. 1990. Livestock marketing and trade in the Mali, Burkina Faso–Côte d'Ivoire corridor. — Abt Associates, USAID, Washington.

Lele, U. and Christiansen, R. 1989. Markets, marketing boards and cooperatives in Africa — issues in adjustment policy. — MADIA discussion paper 11, World Bank, Washington.

Maxwell, S. 1990. Food security in developing countries — issues and options for the 1990s. — IDS Bulletin, 21-3.

Monimart, M. 1989. Femmes du Sahel — la désertification au quotidien. — Karthala, Paris.

Platteau, J-P. 1988. The food crisis in Africa — a comparative structural analysis. — Working paper 44, World Institute for Development Economics Research, Helsinki.

Raikes, P. 1988. Modernising hunger–famine, food surplus and farm policy in the EEC and Africa. — CIIR, Currey, Heinemann, London and Portsmouth.

Reardon, T. 1989. Cereal demand in West Africa — implications for Sahelian regional protection. Paper presented at the Club du Sahel, CILSS seminar on regional cereals markets in West Africa, Lomé.

Richards, P. 1985. Indigenous agricultural revolution. — Hutchinson, London.

Snrech, S. 1988. La dynamique d'organisation du monde rural Sahelien. — D(88)325, Club du Sahel, CILSS, OECD, Paris.

Staatz, J., Dioné, J. and Dembélé, N. 1989. Cereals market liberalisation in Mali. — World Development, 17-5.

Streeten, P. 1987. What price food? — Agricultural price policies in developing countries. — Macmillan, London.

Toulmin, C. and Chambers, R. 1990. Farmer first — achieving sustainable dryland development in Africa. — Issues paper 19, Dryland Network, IIED, London.

Watkins, K. 1991. Agriculture and food security in the GATT Uruguay Round. — Review of African Political Economy, 50.

World Bank 1986. Poverty and hunger — issues and options for food security in developing countries. — World Bank, Washington.

The Sustainability Concept in Relation to Rural Development: Offering Stones for Bread[1]

Henk Breman
University of Wageningen, the Netherlands

De Wit *versus* Pavlov

During a recent visit to Gorom Gorom, in northern Burkina Faso, I tried to verify my impressions of over-grazed rangelands by inquiring into the rate of animal production. Cattle raisers, being farmers in the first place, told me that heifers have their first calf at age of five years. By the time they are 10 years old, the total number of calves is only three. At the end of the dry season it often happens that one has to force cattle to go grazing by pulling them up by their tails.

Similarly, on very poor eolian sands at the western border of Germany the farmers in the past called their cattle "Schwanz Vieh" at the end of the winter. On the other side of the border, in the Dutch province called Drenthe, the situation was similar. And as in Gorom Gorom, the bad performance of cattle was accompanied by severe desertification.

Desertification control started to be successful in Drenthe during the First World War, when the prices for dairy on the European market permitted the use of chemical fertilizers on rangeland, while their use in arable farming caused a rapid decrease in the need for manure. Over-grazing ceased and rural development brought prosperity.

The prosperity is threatened today by the saviour of yesterday, the chemical fertilizers. On the poor sandy soils of the Netherlands we have an annual remainder of nitrogen alone between 150 and 1100 kg/ha depending of the land use: arable farming exclusively or intensive animal farming alone. As we can afford the luxury to care about our environment, research institutes, like the Centre for Agro-biological Research (CABO-DLO) concentrate now on the search for sustainable production systems, which discourage the use of fertilizers.

Most western countries encounter the same problems. Chemical fertilizer is becoming more and more of a negative word. Not only in

---

[1] The present paper is a composition on the three following papers (Breman 1990a, 1990b & 1992), in which further documentation and references can be found. These papers are available on request

relation to the intensive agriculture of the rich part of the world, but also in the development aid community. The latter argues that also the economic sustainability is threatened, when rural development of the poor is based on fertilizers, creating permanent dependence on the rich.

I do not know if you or Danida show this conditioned reflex, this reaction of Pavlov, hearing the word chemical fertilizer. In that case I hope to convince you that a careful system analysis, based on the theoretical production ecology of De Wit is preferable if you care for Third World development.

Analysis of agriculture

Method. — The degree to which environmental factors limit an increase of agricultural production and rural development can be determined by an analysis which compares the production potentials of the natural ecosystem along with the actual agricultural production and by analyzing the possible difference between production potential and actual production.

Agro-ecological research enables determination of the production level agriculture that can be achieved without impairing the carrying capacity of the environment and of the factors quantitatively characterizing this carrying capacity.

Farming systems research provides insight at the level of actual production and into the factors determining this level. A comparison of this level with the theoretical maximum obtainable production, with regard to the environmental carrying capacity, leads to an initial characterization of the situation: a) underexploitation of natural resources, b) actual production equals the maximum production with respect to the carrying capacity, c) over-exploitation of natural resources. In this case, production is higher than that which environmental carrying capacity can cope with and inroads are being made into the available natural resources, instead of profit being made from the early interest, and d) production exceeds the potential of natural resources. Here, external inputs are introduced to alleviate limiting factors of the environment and to upgrade its carrying capacity. In our own modern agriculture, the intensive implementation of these external means reaches the extent, where they become a significant cost factor and entail substantial modifications in natural resources and production methods.

In order to ascertain whether the constraints for production increases are of a social, economic, technical or ecological nature, further specification is necessary.

The under-exploitation of natural resources can originate from: a1) a certain lifestyle, a2) ignorance, or a3) social inequality. Matching production potential of the natural resources can indicate an optimal use of resources b1). With sub-optimal use b2) there is the question of some over-exploitation, it is partly a form of c). Over-exploitation can be a matter of: c1) ignorance, c2) obstinacy (doomsday mentality) or c3) necessity (over-population). With intensive use of external inputs d), specification of the sort and amount is desirable. It can then be estimated to what degree production limitations in the natural environment have been eliminated and which ecological limits apply to the new carrying capacity level. Have water shortages (irrigation), shortages of nutrients (fertilizers, concentrates) or constraints of weeds, diseases and plagues (pesticides) been eliminated?

Unique ecosystems. — Production of plant material is the basis for both arable farming and animal husbandry. As long as man depends almost exclusively on renewable resources, the level of production is determined by the interaction between climatic conditions, soil and plant properties. The plant properties themselves depend strongly on the first two factors. In the long run the vegetation also influences the soil.

The climatic conditions in the Sahelian countries are unique. The sequence of climates prevailing in the Sahelian countries is found in only one other place: along the Australian coast, from the extreme west to the north-east. The influence of the sea, however, and the greater distance from the equator imply that with equal rainfall, the temperatures in tropical Australia are lower, while rainfall becomes less seasonal at increasing aridity. The potential evapo-transpiration (PET) for vast regions of West Africa is, therefore, at least 1000 mm/year higher! Even at 1400 mm annual rainfall, in the West African Guinean savanna, the PET value of 2200 mm/year exceeds still with 450 mm the value for the extreme north of Australia, a region with the same amount of rainfall. The extreme combination of concentrated monomodal summer rainfall, high temperatures during the growing season and very high temperatures combined with very low atmospheric humidities in the dry season, exists nowhere outside the Sahel.

The soils are less typical, but three dominating characteristics augment the consequences of the climate on the vegetation. Precambrian shields form the bed-rock for most soils. As a consequence soils are old, leached and poor. The drier parts of the region are dominated by deep sandy soils, thus the wetting by rainfall is superficial and homogeneous so the evaporation losses are high. Further south with a higher rainfall loamy soils dominate, which are often shallow and easily crusted. The water losses by run-off are great and the water storage capacity is low.

Annual plant species are very competitive in semi-arid climates with one reliable growing season and extreme aridity during the rest of the year. The Sahelian climate is the most typical example, its aridity being accentuated by the low water storage of the soil. As a consequence, the climax vegetations in the Sahelian countries are dominated by annual plants with a short growing cycle, especially in the actual Sahel. If perennial species are present, they are sensitive to disturbance; this is even true for the Sudanian savanna with less harsh conditions.

Annual plants, especially those with a short growing cycle on poor soils, have little impact on the soil: the surface protection is low, the accumulation of plant nutrients in the upper part of the soil profile is limited, as well as the total accumulation of organic matter. That is why even the natural, unexploited ecosystems show a very low availability of nutrients for plant growth. The most limiting element is nitrogen, followed directly by phosphorus. For the annual growth of the herb layer, 7–35 kg/ha of nitrogen is available in the transition zone from the Sahel to savanna. This maximum supposes no losses due to bush fires, herbivores and insects. Phosphorus availability fluctuates around a value of about one-tenth of the nitrogen availability.

Less nutrients are available in agro-ecosystems without external inputs. A certain increase in losses is unavoidable exploiting the ecosystems; the presence of perennial species has still to decrease, to avoid competition. For stable rangelands in the transition zone, the annual nitrogen availability for plant growth is about 15 kg/ha. For arable farming the final stable level will be less than 10 kg/ha. The consumable amount of nitrogen for man at sustainable production is less than 0.5 kg/ha in the first case (as animal protein), and less than 5 kg/ha in the second (as cereals).

These values are even lower than those reported for the semi-desert of Patagonia and for the northern Negev desert.

North–south gradient. — In West Africa, a strong gradient exists in precipitation, which increases from the Sahara in the north, via the Sahel and the savanna, to the tropical rainforest. The variability in rainfall decreases with increasing annual rainfall. A dry year with a probability of 10% (9 out of 10 years will be more humid) has only half the rainfall of an average year at 200 mm, 65% at 400 mm and 80% at 1000 mm.

The isohyets run roughly parallel to the latitude. As a consequence, the length of the growing cycle of most plant species decreases with increasing latitude. The annual species of the region are photosensitive, *i.e.,* daylength determines the length of the growing cycle. This contributes to the indicated advantage for annuals in their competition with perennials.

There are two general clues to the competition between annual plant species on the one hand and perennials (grasses, shrubs, trees, *etc.*) on the other: competition for water determines the abundance in the region, and competition for nutrients and light determines the success in producing biomass. In the northern Sahel, annuals dominate and perennial grasses and woody species are almost absent. The dominance decreases with the increase of water storage outside the reach of annuals: in the southern Sahel, where light soils becomes less dominant and redistribution of rainwater becomes more general, deep water infiltration may occur. The increased abundance of perennials in such areas is often not accompanied by a visible decrease in productivity of the annuals, because the availability of nutrients increases as well as the availability of water. The productivity of annuals decreases when the overall importance of woody perennials increases in the savanna. The increasing cover and the associated annual production from the woody species are associated with a decrease in the availability of nutrients for the herb layer. This is still masked in the Sudanian savanna where perennial grasses gradually replace the annual plant species going south, because perennials are able to produce more biomass at the same nutrient availability, as long as they are not intensively exploited.

Perennial grasses, shrubs and trees extend their habitat to the north, like annual plant species with a long growing cycle, during long periods of good rainfall and relative low exploitation intensity of the vegetation. But a few years of drought are enough to push them back again, far to the south. This is a much faster process than the extension of habitats to the north, which is counteracted by the predominantly southward seed dispersal mechanisms: the Harmattan, the dry wind from the Sahara, and dispersion by migrating herds.

The supply of plant nutrients from natural sources is low across the entire north–south transect. Nutrient availability increases, however, slightly with increasing rainfall due to direct and indirect effects. For example, average annual nitrogen uptake by the above-ground plant parts of the herb layer in natural rangelands increases from 2.5–12.5–20 kg/ha with annual precipitation increasing from 100–400–1000 mm, respectively. At higher rainfall the availability for the herb layer decreases, while the total availability for plant growth still increases, with an increasing proportion being monopolized by the woody species. The nutrient availability for sustainable arable farming will increase proportionally to that of the herb layer, but the absolute values will be lower. The elimination of part of the cover of woody species will appear positive for a while, by the elimination of the competition. But soon the lack of maintenance of the organic matter pool of the soil and the increased leaching become visible. The low biomass of crops and the lack of perennity are accompanied by increased losses of nutrients at increasing rainfall, cancelling the positive influence of the elimination of woody species.

Beside the absolute availabilities of water and nutrients, their relative availability in view of the need for plant growth are important. At less than 200–250 mm of annual infiltration, water is the most limiting factor. This is in the northern Sahel. Further south, the availability of nutrients is more limiting, except in those places where strong run-off and/or shallow soils limit water availability to such values as are found in the north.

Land use. — The proportion of water to nutrient availability has important consequences for the quantity and quality of rangeland production. The nutrient content and hence the feeding value of the herb layer is high if moisture availability is relatively low, so that nutrients cannot be diluted, *i.e.*, in the northern Sahel and on sites with strong run-off. Further south, plant production increases more than the nutrient availability at increasing moisture availability. Since that is associated with increasing nutrient stress, the quality of the herb layer decreases. The negative consequence for herbivores is partly counteracted by the increasing length of the growing season and the increasing cover of woody species, favouring selection by the animals.

The significance for the animal production potential is illustrated by the potential annual net weight gain of heifers in relation to the average rainfall. The weight gain is highest in the north and lowest in the southern Sahel. Going further south the weight gain increases slowly, thanks to the longer growing season and the increasing abundance of perennials.

In most of the southern Sahel and the Sudanian savanna the weight gain is insufficient to guarantee viable production systems: the age at first calving and the calving rate are simply too low. Sustainable animal husbandry is only possible in those regions, as long as no fertilizers are used to improve the availability of high quality fodder, if: a) the region is used by mobile systems in close linkage with the use of the rangelands of the north Sahel, and/or b) only a small fraction of the overall fodder production is used, in other words, if the stocking rate stays very low.

The current production systems have developed in close harmony with the potentials of the ecosystems. The high quality of the water limited rangelands of the northern Sahel and the low quality of rangelands with nutrient-limited production elsewhere have been crucial. The semi-nomadic trans-humance, for a long time by far the most important production system, takes advantage of the high quality fodder of the north during the rainy season. But the fodder production is low and permanent drinking water is scarce outside this season so most pastoralists try to sustain the animals around permanent water in the south during the dry season, preferentially at places with stands of perennial grasses, showing some growth. Animals in the nomadic system using the northern rangelands throughout the year are limited just as the sedentary systems in the south have been limited for a long time. Although the quality of the rangelands increases slowly from the southern Sahel towards the rainforest zone, the increasing humidity also aggravates the problem of sleeping sickness.

Arable farming. — The monomodal distribution of rainfall in a certain season, favouring the occurrence of annual plant species, is favourable for arable subsistence farming. Cereals are the obvious crops in view of the quantity and quality to produce per unit of area.

The northern savanna has been the main centre of development for arable farming: a) the rainfall is high enough, while leaching is still modest and the variability of rainfall is rather limited, and b) the availability of nutrients for growth of annuals is at a maximum.

Preferably the soils are being used the deep loamy sands and sandy loams, which have a high water storage capacity and are relatively easy to farm. Cropping is less frequent in the southern savanna, with heavily leached soils and a higher pest and disease pressure. The actual Sahel, with the much higher risk of low rainfall, is even less attractive for arable farming.

In other words, the northern Sahel is the exclusive domain of animal husbandry; water is the most limiting factor. Further south nutrients are more limiting. These areas are more suitable for arable farming. The highest potential is reached just south of the region with the lowest potential for animal husbandry. In spite of the high rainfall, the Guinean savanna further to the south is also less suitable for arable farming as leaching and water erosion are serious problems. An advantage in comparison with the two former zones is the increased suitability for animal husbandry, provided that diseases are controlled.

Mixed cropping became a way to spread the risks such as diseases, plagues and unfavourable weather conditions. Sorghum, millet and maize are mixed with slow-germinating, nitrogen fixing leguminous species like peanut, cowpea and bambara groundnut ("vouandzou"), *Hibiscus* species yielding fibres and fruits, *etc.*

Mixed cropping systems take the maximum profit out of niche differentiation coping with the limited availability of nutrients. A special form is agro-forestry, for centuries practised in the region, saving species like *néré (Parkia biglobosa)* and "buttertree" (*Vitellaria paradoxa*) at bush clearings. The best known example is crop production in the *Faidherbia albida* parkland. Those systems, being far from effective enough to sustain crop yields in view of the characteristics of the ecosystems, are combined with the practice of fallow far from the village and the use of manure around the village.

Development context

Introduction. — By translating the annual productivity of the natural ecosystems into productivity of agricultural systems one may estimate the carrying capacity of agro-ecosystems in the region. The most accurate translation is that based on the limiting factors for primary production: water in the north, nutrients elsewhere. If one is aiming at sustainable use of the renewable resources, availability of that limiting factor should not decrease.

The availability of the limiting factor for sustainable use is less than the apparent availability, the quantity used for the actual overall annual production. The reason is the involvement of stocks, built during thousands of years with low exploitation pressure: deep ground-water, nutrients and organic matter in the soil profile and organic matter in perennial plants, particularly shrubs and trees. Without the use of external inputs, man draws more from those stocks through arable farming than through animal husbandry. The output of the latter approaches the "consumable nutrients" at a sustainable basis better than the first; the consumable output of nutrients from crops needs to be corrected for the fraction representing resource exhaustion. The best possible estimations are based on a careful analysis of the water and nutrient balance. The available analyses are more detailed for the Sahel than for the savanna, but the synthesis of agricultural research in francophone Africa completes the overall picture.

It is easy to translate "consumable nutrients" into animal or crop production. But to translate these into the carrying capacity for man becomes difficult as soon as the consumption is not direct, *i.e.,* when markets and prices are involved. Here, only production for local subsistence has been taken into account as far as crop production is concerned. In case of animal production, the exchange rate between milk and cereals and the producer prices for animals from the 1970s are used, when the EEC and Argentina had not yet taken over the coastal markets of West Africa. For the zone where nutrients are the limiting factor, the capacity of the agro-ecosystem to feed man will not only be presented for arable farming. The capacity based on animal production is needed too because: a) part of the land cannot or should not be used for crop production, and b) animals are used to support arable farming.

The carrying capacity will be expressed in number of people per square kilometre, the maximum population density based on the renewable resources only. For the Sahel this figure will be based on the production in a dry year, defined as a year with a probability of occurence of 10%. The maximum population density will be compared with the actual population density and the consequences of over-population will be discussed.

Sustainable yields of animal husbandry. — The level of production in animal husbandry is a function of the inversely proportional stocking rate and individual performance. Taking both aspects into account, the animal protein production has been estimated per zone for pastoral systems and for agricultural systems, using animals. For the northern Sahel only pastoral production has been estimated, but pastoral as well as agricultural animal production were estimated for the other zones. While in the Guinean savanna sedentary pastoral production is most important, mobile trans-humance predominates in the southern Sahel and the Sudanian savanna, linking both zones to the northern Sahel.

Table 1 presents the sustainable annual protein production. Protein mass was chosen because of its relevance for human nutrition and the direct link with the "consumable nitrogen". Protein output equals 6.25 times the "consumable nitrogen" output. The pastoral production goal leads to systems with a comparative high production, trans-humance being the most productive.

Manure and energy for traction are important products particularly in the sedentary agricultural systems. The maximum amount that theoretically will be available is discounted in the case of integrated resource use.

The manure production is estimated on the basis of stocking rates of 22 Tropical Livestock Units (TLU)/km^2 for the trans-humance and for the sedentary systems in the Guinean savanna, and of 20 TLU/km^2 for sedentary systems elsewhere. Only 35% of the daily manure production of trans-humance herds can be collected during eight months per year at most. This yields about 5,000 kg/km^2 of dry manure. In the other systems 50% of the annual production can be collected at most, or 10,000 kg/km^2 for the Sahel and the Sudanian savanna, and 11,000 kg/km^2 for the Guinean savanna.

Sustainable yields of arable farming. — The sustainable cereal yield of arable land in the Sudanian savanna, which is the most suitable region for crop production, is 300 kg/ha. Comparing this yield and the required amount of consumable and available nitrogen with the variation of available nitrogen caused by rainfall variability, the sustainable yield for both other zones with arable farming was calculated. The average yields in the southern Sahel will be 190 kg/ha, but in a dry year it will, at most, be 120 kg/ha. The Guinean savanna will not produce the same amount of cereals as the Sudanian savanna on a sustainable basis. If the food production, however, is mainly realized by perennial crops, while woody species are saved as much as possible on the fields, it is possible to reach a food production equivalent to at most 300 kg/ha of cereals.

Estimating the fraction of land suitable for permanent cropping at 25, 35 and 50% for the southern Sahel, Sudanian and Guinean savanna, respectively, cereal productivity will be 2,400, 8,400 and 12,000 kg/km^2 for the three zones respectively. Harvest and storage losses are supposed to be 20%.

System	Zone	Goal kg/km^2/year	
		pastoral	agricultural
Nomadism	northern Sahel	50	—
Transhumance	Sahel/northern savanna	340	—
Sedentary	southern Sahel/Sudanian	—	70
	Guinean savanna	140	90

Table 1. The protein production (kg/km^2/year) of different animal production systems in different agro-ecological zones.

Sustainable exploitation of wood. — The acreage of natural rangelands required to cover the need of wood has been estimated in the beginning of the 1980s. This was requirements of about 4, 1.6 and 0.6 ha/capita in the southern Sahel, in the Sudanian and Guinean savanna, respectively. In view of the high mortality of shrubs and trees during two decades of drought, the need in the Sahel will be 8 ha/capita today.

Carrying capacity. — The above data permits the estimation of the carrying capacity. It is supposed that an absolute minimum of either 250 kg of cereals per head is needed, or the output of three TLU/capita, at the productivity level of the mobile systems. The number of TLU/capita for the other systems is linearly related with their protein production (Tab. 1).

The carrying capacity of the various zones is presented in Table 2 for the different systems. In the case of integrated land use the maximum amount of manure is collected and used to increase cereal production. It is estimated that 1,000 kg of manure increases cereal output by 150 kg of grain at the most. The carrying capacity under integrated land-use is not simply the sum of the values for exclusive use by pastoral systems or arable farming. It is supposed that arable farming will occupy the best soils, in other words the acreage available for animal husbandry decreases, as well as the rangeland productivity.

The production of wood is almost as limiting in the Sahel as the agricultural potential since the drought. This is calculated from a comparison of the acreage per person in relation to the carrying capacity for integrated land use, which represents the maximum sustainable population density, with the acreage of rangeland needed to satisfy the wood production. If 25% of the zone is occupied by fields, the rest produces enough wood for about 10 persons/km². Elsewhere the potential food production is much more limiting than wood production for sustainable domestic use. However, this supposes a homogeneous

Zone	Carrying capacity persons/km²			Population density
	animal husbandry	arable farming	integrated land use	actual land use
Northern Sahel	1 (nomadic)	—	1	1 (0–7)
Southern Sahel	7 (transhumance)	10	11	13 (7–27)
Soudanian savanna	7 (transhumance)	34	36	33 (7–66)
Guinean savanna	3 (sedentary)	48	51	25*

Table 2. Carrying capacity of the agro-ecosystems (persons/km²) in relation to land use, in comparison with the population density. * Based on data from Mali and Burkina Faso only.

distribution of the population.

A comparison between the actual population density and the carrying capacity for integrated land use suggests a saturated northern Sahel, a seriously overpopulated southern Sahel and an almost saturated Sudanian savanna, locally heavily over-populated. The abundance of land is fictitious, absolute population density by itself is no good criterion, but should be interpreted in relation to the potentials of the natural resources. Less alarming appears to be the situation in the Guinean zone, if at least the assumption is realistic that nutrient losses are limited in perennial crops and in agro-forestry. The difference between the actual population densities of the Sudanian and the Guinean zones is not necessarily due to diseases such as sleeping sickness, river blindness and malaria alone. Maybe the lower suitability for arable farming is to be blamed as well.

The analysis leads to the conclusion that the main problem of the Sahelian countries is not the drought, but the over-exploitation of the natural resources, which are naturally poor. Over-population will be the main reason for over-exploitation, over-population at a relatively low population density (Tab. 2).

The fact that part of the population has sources of income outside agriculture, and part of the agricultural production is based on external inputs, does not play a major role. Non-renewable resources are rare and the agricultural production based on the use of chemical fertilizers represents only a small percentage of the total production. Neither lack of knowledge nor unequal access to the natural resources are the ultimate reasons for the over-exploitation of these resources, although their influence increases due to the deterioration of the former production systems and of the natural and social environment.

Recent developments. — The typical climate of the region favours arable farming at an aridity that would elsewhere only permit wildlife or animal husbandry. As a consequence, the population density can rise easily (Tab. 2), while the buffering stocks of above-ground and soil organic matter is lower. This summarizes why the agro-ecosystems of the region are so fragile.

Well-designed production systems, based on the ecology of the region, and a high growth rate of the population have caused over-population. The effective transhumance has caused over-population even in zones dominated by animal husbandry.

The low buffer capacity of the ecosystem signifies that over-exploitation resulting from over-population will cause a decrease of natural production potential sooner than elsewhere. Over-population may be masked, temporarily, by a long period of high rainfall. After such periods, severe effects may be expected as have been observed. Not only the environment breaks down, but the well-adapted production systems too. The migration pattern was not calculated for the extreme drought of 1972 and 1973 and both herdsmen and herds died. To save their animals farmers had to cut down their *Faidherbia albida* trees. Crop species and varieties with a growing cycle calculated for "normal" dry years, needed a longer growing season than the rainfall of 1972 and 1973 offered, causing much lower yields than necessary.

In the northern Sahel over-exploitation of the environment is mainly caused by overgrazing. In the southern Sahel and the Sudanian savanna arable farming is also a cause and wood cutting should not be ignored either.

The most severe direct effect of overgrazing is decreased water infiltration (increased run-off). It affects in particular the loamy soils, *i.e.*, half of the soils of the southern Sahel and most of the soils of the savanna. Soil slaking and crusting cause lower water availability, but higher fodder quality. As a consequence, once started, overgrazing is difficult to stop. The most serious direct danger of too intensive arable farming is exhaustion of the soil; too intensive wood-cutting leads to destruction of the production capital. Effects of all three forms of over-exploitation are wind erosion (Sahel) and water erosion (savanna and southern Sahel). In the case of animal husbandry in the southern savanna, erosion is locally hampered by bush encroachment, but the net result is that fodder availability decreases.

Only locally has the degree of the resource degradation been quantified. The results indicate that the production potential based on an uneroded environment, like above, causes serious over-estimation. The carrying capacity will be less than Table 2 suggests. On the other hand, not the entire population of the region relies on the exploitation of renewable resources.

Over-population and drought are not the exclusive explanations for recent developments in agriculture. Since the colonial era, cash crops were introduced under pressure, boreholes were made, veterinarians became active and land tenure was changed. The rights obtained by land-clearing were recognized, but pastoral regulations were ignored. These matters reinforced the adverse effects of over-population, drought, and the increased competition for land, with farmers as the overall winners.

The loss of dry-season pastures and the bad exchange rate between animal products and cereals for pastoralists during the drought meant that many animals fell into the hands of farmers and speculators. Farmers need cattle to sustain crop production, speculators try to continue the former pastoral production, in view of the relatively high rate of return on the investment, or to start semi-intensive milk production systems around cities. The former pastoralists are obliged to serve as herdsmen for the new owners or to settle as farmer or agro-pastoralist. Many of those who saved their herds are now occupying areas in the extreme south of the savanna, with low population pressure.

The main focus of animal husbandry moved southwards, to zones with lower production potential. Population groups that used to be complementary, in terms of their production systems, are now increasingly competitive. The overall food production is less balanced and an important source of foreign currency is degrading.

The most important change in arable farming is the intensification of cotton production. In the past this cash crop caused pressure on the production of food crops, today the cereal production increases too, albeit slowly. This does not signify, however, an end to soil depletion; the losses are still increasing. An important reason is a new linkage between arable farmers and stock farmers. Where trans-humance herds used to bring fertility to the farms through manure contracts, farmers nowadays "sell their soil fertility" to semi-intensive animal production systems, like those producing milk around the cities, in the form of cotton cakes.

Sustainable production

Options to reduce over-exploitation by over-population are migration, alternative employment and intensifying agricultural production. Inter-zonal migration alleviates over-exploitation slightly, but socio-political and economic considerations impose strict limits to the volume of migration. International migration could be effective, but is completely theoretical. Even the Brundlandt report speaks only about regional equity, to say nothing about the EEC agreement of Schengen.

Prospects for an increase of employment outside agriculture are limited. It is, however, worthwhile to consider it in development planning, by comparing the cost/benefit relationship with those of the alternatives, intensifying agriculture or permanent aid. In particular, when environmental damage, social insecurity and unrest are discounted as costs, there will be occasions where the comparison favours *e.g.,* local processing of agricultural products or import replacing industries. The local poisoning of the environment by dirty industries may be preferable above its general destruction by over-exploitation.

Intensification of agriculture has the highest potential to stop over-exploitation. With nutrients being the limiting factor in the zones suitable for arable farming, improvement of soil fertility is theoretically the weapon to fight over-population, unless alternative employment is cheaper to create. Even in the southern Sahel the rangeland production could be five-fold with unlimited availability of nutrients. The same is true for cereal production in that zone, and southward into the savanna the effect of improving soil fertility still increases. Leguminous species are able to amend the lack of nitrogen, the most limiting nutrient. If enough phosphorus is available, even the indigenous rangeland species are able to increase the nitrogen availability, and thus animal production, by one third in the Sahel. For the savanna a more than ten-fold increase is conceivable, in view of the experience in the extreme north of tropical Australia. For leguminous food crops the same has been shown.

Technical options to solve the problem of nutrients shortage can be divided into two groups, those based on internal inputs alone (manure, biological nitrogen fixation and agro-forestry) and those which need the use of chemical fertilizers and other external inputs For a region like the Sahel, over-exploited by over-population, the first group does not offer a solution any more. Only Baron von Münchhausen was able to pull himself and his horse by its own hairs out of the moor! All potential available manure has been discounted in the establishment of the maximum population density at integrated resource use (Tab. 2). The efficient production systems of the past came into trouble in spite of their intensive use of leguminous species in mixed cropping and agro-forestry.

The English expression "run with the hare and hunt with the hounds" indicates that man continuously tries to combine conflicting things, which in Dutch is "save the cabbage and the goat." In relation to the Sahel I should say: in case of over-population it is impossible "to save the *Acacia* and the goat."

In other words, the use of chemical fertilizers is the only option able to trigger rural development, to assure sustainable production. As a consequence food security is unattainable if the cultivation of commercial crops is ignored

But cotton seems to be the only important crop on which its use is economical attractive, particularly in the southern part of the Sudanian savanna. Other cash crops for which the region has a real comparative advantage have not been discovered yet. The bad cost/benefit-relations are partly caused by limited experience, but certainly also by the low population density: the creation of an effective infrastructure is very expensive. Besides, the extreme climate and the related low availability of organic matter make the use of chemical fertilizers risky, in view of the low cation exchange capacity of the soils. This risk is, however, not

production systems	soil		climate
	fertility	water-holding cap.	water availability
	low high	low high	low high
intensive agriculture			
integrated agriculture			

Figure 1. The production efficiency of intensive agriculture in relation to the potential of the environment, specified for soil and climate parameters, compared witth such a relation for integrated agriculture.

insurmountable.

The question "high or low use of external inputs" should not preoccupy those who worry about sustainable Sahelian production, but "how to maximize the efficient use of external resources?" The real bottle-neck of the agro-ecosystems of the region, the Gordian knot to unravel, is the low absolute population density at which over-population is reached. This makes investments aimed at intensification very expensive, while ecological alternatives are very labour-intensive.

Integrated agriculture. — When trying to develop methods to intensify agriculture in sub-optimum zones, it is dangerously naive to copy the western example. The latter developed without serious incentives for

Production system	Soil						Climate		
	fertility			**water-holding capacity**			**water availability**		
	low	medium	high	low	medium	high	low	medium	high
Increased availability of water and/or nutrients									
Manure		+		+					+
Legumes	+				+			+	
Agro-forestry	+				+		+		
Decreased losses of water and/or nutrients									
Agro-forestry		+		+					+
Wind erosion control		+		+			+		
Water erosion control		+			+				+
Water harvest		+		+			+		
Mixed cropping, pest control		+			+				+
Mixed cropping, drought	+			+				+	
Adapted varieties	+			+			+		+

Table 3. Indications of environmental condition; that guarantee optimum efficiency for agricultural elements regarded as "natural alternatives" for intencification through the use of external inputs.

economizing on the use of inputs such as fertilizers and pesticides and without general awareness concerning the environment. But disregarding the production potentials of the western example would also be unwise, in view of the still growing demands for food. Mixtures of elements of intensive agriculture and "natural alternatives" might lead to a more efficient use of external inputs, more feasible from the economic point of view, with less losses to the environment.

The suitability of a region for intensive agriculture depends on soil fertility, soil water-holding capacity and water availability as dictated by climate. Their rough relation with the efficiency of intensive agriculture is presented in Figure 1. The optimum conditions for natural alternatives are not necessarily identical to those for intensive agriculture. Some of the advantages in Table 3 are self-evident, other were chosen rather subjectively and could probably be questioned by specialists. Given that for some of the conditions the optimum for natural alternatives is different from that for intensive agriculture, combinations of both, so-called integrated agriculture, may have optimum conditions different from those for intensive agriculture alone. At least the range of conditions under which potential production can be realized is extended (Fig. 1). The Malian-Dutch research project "Production Sudano-Sahél-ienne" is developing such an integrated agricultureal system.

Developing integrated agriculture is far from being a agro-technical affair alone. Changing policies of the national government are indispensable in the fields of land use regulations, taxes, credits, subsidies, export, family planning, *etc.*

But instead of thinking about the obligations of the Sahelian governments it is more effective to think about ourselves, about our policy. Instead of offering the Sahelian countries the space on the world market they need to be able to procure the indispensable external inputs, we continue to compete with them on their traditional export markets and even on their domestic markets. A modification of their own agricultural and trade policies is the best help that rich countries can give to create favourable conditions for agricultural development in Africa.

Without such modifications we oblige the over-populated Sahel to continue to intensify their exhaustive production systems. Without such modifications our sustainability concept is only offering stones for bread!

Litterature cited

Breman, H. 1990a. No sustainability without external inputs. — Pp. 124–133 in: Beyond Adjustment Africa Seminar. — Ministry of Foreign Affairs, The Hague.
Breman, H. 1990b. Integrating crops and livestock in southern Mali: rural development or environmental degradation? — Pp. 277–294 in: Rabbinge, R. *et al.* (eds.), Theoretical Production Ecology: reflections and prospects. — Centre for Agriculturel Publishing and Documentation (PUDOC), Wageningen.
Breman, H. 1992 The renewable resources base for Sahelian development: its limits and potentials. — Royal Tropical Institute, Amsterdam.

Problems of Defining and Evaluating Sustainability of Agricultural Systems

Anette Reenberg and *Kjeld Rasmussen*
University of Copenhagen, Denmark

Introduction

Since the beginning of the 1970s the scientific interest concerning the agricultural systems in the harsh lands of the Sahelian region and their impact on the environment, has been considerable and ever increasing. The focus of interest has been to evaluate the threat of desertification in the semi-arid regions and also the relative importance of various causes, ranging from climatic changes to socio-economic factors leading to an intensive stress on the fragile environment.

The negative environmental impact of intensified agricultural activities, *e.g.*, caused by the rapidly increasing population, is widely acknowledged as a major driving factor behind more or less irreversible changes, often generalized in the term desertification.

This paper will not deal with the discussion of whether desertification is taking place in the Sahel nor with a definition of the concept of desertification — even if both matters might deserve attention. Instead, it will concentrate on the question if, and how, it is possible to evaluate the sustainability of a typical agricultural system in the Sahel — exemplified by a millet production system in th Oudalan Province.

Sustainability and agricultural systems

Since the publication of the Brundtland report the term sustainability has been widely accepted, *e.g.*, in relation to evaluation of the human use of natural resources.

In connection with agricultural systems, the term has been used to characterize both the state and function of the system at a certain time and at a certain development. In the present context, only the former use of the term will be considered.

Sustainability involves both ecological, economic and social aspects, but here mainly the ecological aspects will be discussed:
— what resources should be considered the basis for the agricultural production and

— in which way may possible changes in the resource base be measured and monitored?

In other words, we will discuss the possibilities of setting up an analytical framework and to define relevant and necessary parameters to evaluate the sustainability of the typical Sahelian agricultural system.

The agricultural system in Bidi

In the northern part of the Sahelian region, *e.g.*, where yearly precipitation on average is below 400 mm, the possible variations in agricultural strategies are very limited. Roughly speaking, it is restricted to cultivation of millet, supplemented by a little sorghum in low-lying areas and a few other crops, combined with pastoral exploitation in the shape of livestock owned by the local peasants as well as by nomadic pastoralists.

Even if strategies constituting the agricultural systems vary, due to the local differences in the natural environment and the ethnic

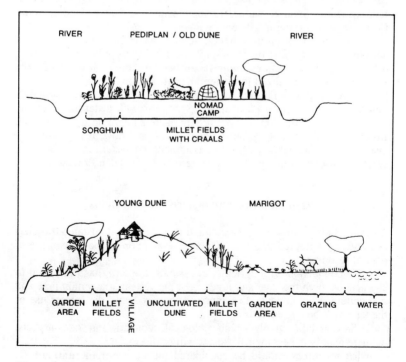

Figure 1. North-south profile of the agricultural system in Bidi.

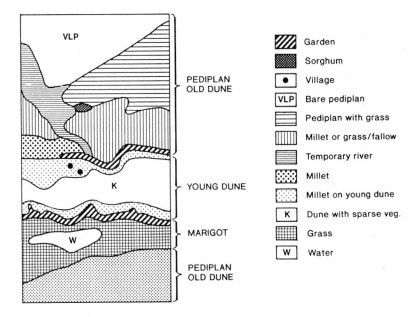

Figure 2. Land-use map corresponding to Figure 1.

composition of the population, it is believed to be an acceptable generalisation to use one local example to illustrate the main structure in the agricultural system used in northern Sahel.

The study area selected for the following discussion on sustainability in relation to agricultural systems in northern Sahel is located in northern Burkina Faso. The environment is "typical Sahelian." The precipitation is highly variable within as well as between years, and the geomorphology and soils can also be termed typical, *e.g.*, older and younger east-west oriented longitudinal fossil dunes alternating with a lateritic pediplain and temporary rivers and lakes.

The village Bidi is located in the Oudalan Province approximately 12 kilometres south-west of Gorom-Gorom. It is, as many other villages in the region, situated on the young dunes and surrounded with a more or less continuous area of fields (Fig. 1, 2). The fields belonging to and cultivated by the Bidi villagers may roughly be divided into seven regions according to land use, productivity and stability.

Fields on the young dune close to the villages Bidi 1 and Bidi 2. — These fields are located around the village on very loose sandy soils. Towards the north the fields are bordered by the gardens at the dune front and, in the other directions, by uncultivated dune with sparse vegetation dominated by grasses and herbs. A gradual change from cultivated to uncultivated dune can be observed, and differences in soils do not seem to be the main factor determining the extension of the fields. This indication is supported by the local peasants, who state that if a villager should need more acreage, larger parts of the dune might be cultivated in the years to come. Especially in the fields that have not been cultivated for some years, accumulation of organic matter is expected to be sufficient to allow for further cultivation in the near future.

Fields on northern side of the young dune in westerly direction. — They are apparently good fertile fields with a high biomass production. The fields are old and have in general been cultivated continuously for many years, but quite a number of random patches with fallow also occur owing to the lack of the necessary man-power for seeding and especially weeding.

The dune-front facing south. — These support a substantial part of the millet fields of Bidi. An approximately 200 m wide band on the dune along the "marigot" is, or has recently been, cultivated. In general, the soils are considered suitable for cultivation, and some very fine fields are located in the region. The field of the *chef du village* has for instance been cultivated for the last 27 years.

Yet, large differences can be observed in the region. The potential outcome from the fields bordering the *marigot* depends on the precipitation level. In years with abundant rainfall it might be impossible to grow millet because seeding is not possible in the flooded fields. Farther away from the *marigot*, the land use pattern ranges from normal yielding to very low yielding fields on "tired" soils and to fallow. The latter will most likely be left uncultivated for some years in the near future. The length of the fallow period is not fixed to a certain number of years, but will be evaluated by "trial-and-error" when some years have gone, the field will be sown, and if the millet germinates properly, the field will be weeded and harvested.

Gardens bordering the marigot towards the south of the dune. — These are used for crops such as sweet potatoes and cowpeas, the two dominating crops, but also minor crops such as okra and sorrel are grown. All crops are cultivated as mono-culture, each occupying a part of the garden. The cowpeas are sown on flat soil, as soon as the water from the marigot is somewhat withdrawn (September in 1990), whereas the sweet potatoes are grown in mounds which are watered by hand, or occasionally by small channels leading water from the *marigot* to the garden. Also in this region the fallow is practised when the soils are "tired."

Gardens at the northern fringe of the dune. — Here a larger variety of crops are cultivated, but again the crops are grown as mono-cultures. They include maize, sorghum, date, mango, maniok, sorrel, sweet potatoe, cotton and tomatoe cultivated for local consumption as well as for the market. The cultivation is intensive and mainly the responsibility of the women and the boys.

The millet field region north-east of the young dune. — These consist of fields of different ages alternating with pastures and fallows. The stability of the land use pattern in this region varies considerably, ranging from continuous cultivation on the most fertile fields for more than 40 years to recent expansions of one or two year old fields, and to fallow. In general, the soils are considered well-suited for millet cultivation and are manured regularly. The stable fields in the region have all been well kept with manure and carefully cultivated for many years. Yet, compared to the young dune, this region constitutes new expansions or displacements of the fields. Old inhabitants in the village can remember the time when only the young dune was cultivated.

The nomads, bellas and peul, have large herds of cattle passing through the area and settling in the fields. The same nomads will often return every year to camps prepared by the peasants. The droppings from the cattle are concentrated by keeping the animals in craals during the night and the location of which is occasionally changed to other parts of the field. The value of the manure is realized by the peasants and the droppings are often supplemented by manure collected outside the field and carried to the field in sacks.

Towards the north, a few, but highly productive sorghum fields border the river. The water nearby favours sorghum cultivation but, in general, millet is preferred as food, and only a limited amount of sorghum is cultivated in the Bidi region.

Also millet fields benefit from being close to the river. A very productive millet field is located in the northern part of the region close to the river where it sustains a small, permanent production unit consisting of one family.

The millet fields to the north of the dune, east of the river. — These are stable old fields cultivated continuously for many years. Generally, they are considered of good quality, benefitting from a relatively high content of organic matter.

Opinions on sustainability of agriculture in the Sahel

In literature dealing with environmental effect of agriculture in the Sahelian region, one often meets one or more of the following statements concerning causes and indicators of the lack of sustainability in the

agricultural systems:
1) intensive cultivation on the sandy soils on the longitudinal dune system has triggered wind erosion with the permanent loss of fertile soil as a consequence (Krings 1980).
2) Increasing population pressure has forced the peasants to cultivate less accessible and more sensitive soil types (*e.g.*, on the pediplain) — with an increasing danger of gully erosion as a consequence (Milleville 1980).
3) Increasing population pressure has led to the disappearance of fallow with a decline in *e.g.*,soil fertility (and yields) as a consequence.
4) The increase in acreage of the cultivated land leaves less space for pastures with the possible consequence that fewer pastoralists are camping in or passing the region; thereby the very essential symbiotic linkage (stubble grazing *versus* manure) peasants and pastoralists is being upset.
5) The local need for fuelwood leads to excessive wood-cutting and destruction of the natural vegetation.
These, and other equivalent, statements have been generally acknowledged, and they certainly affect the reality of the situation. Yet, it might be worthwhile to substantiate whether these statements hold true

Evaluating sustainability of the agricultural system

A simple definition of a sustainable agricultural system could be: a system where production can be continued at the present level without increasing input and without any resulting depletion of the natural resources utilised for the production.

It is a very difficult task, however, to investigate whether this definition holds true or not in an agricultural system such as the one found in Bidi. Even if the keywords in the above definition of sustainability appear quite simple, a precise definition in relation to our specific case raises many questions, some of which are outlined in Table 1. One main reason for the problems is that all parameters needed to describe and define the agricultural conditions, the resource utilisation and the production strategies are highly fluctuating in time and space. Variations from one year to the other are large and a monitoring of possible development trends are therefore very difficult. The most outstanding feature of the system is that all parameters change, often substantially, from one year to the other and that a high degree of flexibility can be observed. Some fluctuations are caused by external factors in both the socio-economic and the natural environment. Some can be considered gradual adjustments in the agricultural practice.

An evaluation of the viability must be based on a thorough understanding of the driving factors behind the choice of agricultural production strategies. Only a holistic description of the agricultural system will provide an adequate framework for an evaluation of the long-term stability. But an evaluation must also stress the fact that, according to the normal definition, a sustainable production does not equal production stability. It is important to envisage that substantial yearly fluctuations in food production will be inevitable and that local food supply will not always be sufficient.

Parameters influencing agricultural decisions

As described above, the peasants in Bidi produce mainly millet supplemented by various crops cultivated in the small gardens and by livestock rearing. The utilisation of the limited natural resources results in a simple land use pattern which in detail, however, is determined by a complex pattern of several parameters:

1) the local population size is an important determinant, acting as a figure defining the need for stable food, as well as maximum man-power available for field work,

2) the fluctuations, long term as well as short term, in the environmental conditions, related to positive and negative effects of water availability, to degradation of soils, *etc.*, and

3) changes in awareness of the potential value of various resources such as *e.g.*, soil types and technology.

Key words	Key questions
Input needed	*Concerning capital:* are earnings outside a normal part of the agricultural system? *Concerning man-power:* is input defined as number of peasants or number of work-hours? *Concerning land:* how is the village delimited?
Output needed	*Concerning cereals:* do we measure yields per ha or total yield fo the village? *Concerning alternative products:* how is their value determined? *Concerning variation:* how is the "output-stability" defined?
Status of the natural resource-base	How is degradation measured? Is reversible degradation considered a normal feature? What is the acceptable time horizon for restoration?

Table 1. Evaluation of sustainability. Key words and corresponding key questions.

Weeding capacity. — Earlier studies from the region carried out by Peretti (1976) and Milleville (1980) discussed in detail key factors important for understanding the agricultural strategy. Both studies stress that the main factor limiting the cultivated area is the man-power available for weeding. A strong correlation (r = 0.80) is found between the weeding capacity, defined as the number of men in the household, as weeding is almost exclusively a male activity, and the acreage of millet by the end of the growing season. The marginal increase in cultivated land to manpower in the study region around Kolel is significantly larger for the villages on sandy soil (dune) compared to pediplain.

The overall increase in the cultivated area was estimated in the study region from aerial photos from 1955 and 1974, with the conclusion that the relative increase was almost equivalent to the population increase (= 60%) and obtained from two sources. One was by expanding the fields on former uncultivated soils in the old dunes or the pediplain, and the other was by a reduction in the fallow.

Seeding capacity. — Even if seeding constitutes a relatively small work input, the actual use of the land might be influenced by the available work force at the outset of the rainy season. The traditional long distance migrations for work may, in case of early rains, imply that the men have not returned in time for the seeding. Also lack of food by the end of the dry season may limit the working capacity. In both cases a relatively large proportion of fallow will be the result.

Manuring capacity. — Manure is another very important production factor. The main sources are animals owned by the local peasants and animals migrating with the nomads, but camping temporarily in the fields for part of the dry season. The manure left might, according to Milleville (1980), reach 10 tons dry matter per ha in parts of the field, a figure which is even considerably lower than estimates made by the present authors in Bidi 1991.

In order to distribute the manure evenly, temporary camps are often established in the field by the peasants and displaced several times during the season. In many fields, however, the manure is distributed by the peasants who collect the manure outside the field and carry it in sacks to the field. The latter mode constitutes an important manpower bottle-neck, and lack of time is often reported as the reason for an insufficient manuring with a resulting substantial decrease in yields on parts of the field. A more or less regular two-year rotation is often practised in order to ensure a proper manuring every second year.

The actual benefit of the manure to the millet depends very much on the availability of water. If water is insufficient, the manure will burn the millet.

Fertility fluctuations. — In addition to the manpower capacity, the soil fertility determines what areas will be under crops in a specific year. Some fields are left fallow when considered "tired." After some years of rest they will be sown, but the length of the fallow period follows no specific rules, and in general it must be recognized that the fallow concept is difficult to handle in this connection, because it is almost impossible to distinguish fields that have not been weeded from real fallow.

Indicators for non-sustainability

On several occasions the region in question has been pointed out as one severely threatened by desertification (Fig. 3) as a result of over-grazing and over-cultivation. The hypothesis behind these statements is based on the existence of vicious circles including:

1) a permanent destruction of pastures owing to overgrazing,
2) an inclusion of new land for cultivation on soil types more sensitive to erosion,
3) a depletion of soil nutrients owing to a lack of fallowing and
4) declining precipitation trends, *etc.*

This hypothesis has been supported by various observations. Krings (1980) pointed to the fact that the former natural vegetation on the young dune has been replaced by millet fields which thus gradually

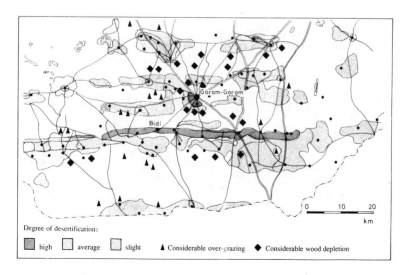

Figure 3. The degree of desertification in the study region as evaluated by Krings (1980).

turned into large areas dominated by loose, moving sand. A change which is considered irreversible. Milleville (1980) investigated the encroachment of millet fields into former pasture land. The increasing population necessitates a need for more cultivated land which is obtained either by a more intensive cultivation (less fallowing) or cultivation of new land. According to Milleville (l.c.), both strategies are believed to cause degradation of the environment in the form of depletion of soil nutrients, or accelerated soil erosion on the pediplain soils.

Yet, the hypothesis concerning the increasing pressure on land and its long term environmental consequences might deserve reconsideration. It needs to be based on a solid, empirical knowledge of changes in land use, in yield, in the composition of the natural vegetation, soil nutrient content and soil erosion indicators.

Figure 4. Aerial photo from Bidi (1:50,000 from 1981). A, B, C and D indicates the location of the fields presented in Table 2. Examples on the "history of cultivation" are roughly indicated for the ares within the dotted line as follows: n = fields taken into cultivation within the last 40 years, c = areas on the dune still cultivated, x = fields abandoned 6–8 years ago, y = fields abandoned more than 15 years ago, z = fields abandoned 1990.

Experience from the Bidi-case

A few preliminary results of investigations in Bidi carried out in 1990 and 1991 might illustrate potential alternative interpretations of the observed changes in important indicators.

Alterations in land-use versus *soil types.* — Within the last 20 years the locations of the fields cultivated by the peasants in Bidi have gradually changed. From being almost exclusively located in the young dune, the fields have slowly spread to the pediplain towards the north.
But the new expansions have been followed by an almost equivalent abandonment of the fields in the dune (Fig. 4).

In contrast to what was expected, the reason given by the local peasants was that the pediplain had proven to produce higher yields. The decisions concerning site selection for the fields are apparertly determined by very individual and subjective priorities, based on former experience, importance of manpower bottle-necks for weeding, *etc.* (Tab. 2).

Yield changes. — The above mentioned fields on the dune have been abandoned after years with very low yields. They needed fallowing to regain fertility, but are, at least by some informants, evaluated as suitable for recultivation after some years because of sufficient accumulation of organic matter. Others regard the fertility change as caused by a permanent loss of the fertile topsoil. Evaluation of these opposite statements may be difficult due to the fact that no detailed registration is available of soil fertility in, *e.g.*, the 1970s. But the question is also a reminder of the importance of defining the time scale we are considering when we try to define sustainability. How much time will we allow for a natural regeneration of the resource base?

	Size in ha	Yield, bundles 1991–1990	Yield, kg/ha 1991–1990
A. Pediplain	3,4	60/ 3	194/ 97
B. Dune	1,4	13/ 7	102/ 55
C. Pediplain	4,0	130/ 8	358/ 22
D. Dune	0,8	25/ 4	344/ 55

Table 2. Yield estimates for four selected fields in the Bidi region (locations indicated on aerial photo in Fig. 4). 1991 figures represent extremely high level caused by abundant precipitation. The estimates are based on farm interviews (number of bundles harvested) and a sample based average weight for the millet bundle. Fields A and B belong to one household and C and D another. See text for a discussion on the yield level for various parts of the landscape and their possible influence on the land use strategy.

Soil nutrients. — Results from a preliminary soil survey in 1990 indicate that the availability of nutrients (especially N and P) in general is very limited. This implies that manure probably constitutes an important part of the nutrient circulation budget. If so, the availability of manure will be a dominant parameter for an evaluation of the sustainability.

Regrowth of the natural vegetation. — Observations performed in 1990 and 1991 in the non-cultivated parts of the young dune reveal a considerable regrowth of the natural vegetation on the former fields, presumably favoured by the abundant rains in 1991. There seems to be no immediate evidence of a permanent deterioration, yet, it has to be investigated whether possible changes in vegetation composition should indicate some inherent degradation of the environment.

Conclusions

In the discussion of the sustainability of the Sahelian agricultural systems, it must be recognized that the traditional and locally accepted living conditions for the rural population in the Sahel, exemplified in this presentation by the village Bidi in Oudalan, are based on social and economic norms very different rom those found in, *e.g.*, the European agriculture.

The agricultural production highly much dependent on the variable rainfall and is sufficient to cover local food requirements only in good years. The deficit is covered through salaries obtained from traditional long distance migrations for work during the dry season, when a large percentage of the men are working in the gold-mines or in big cities in the neighbouring countries.

In spite of the fact that the economic value of the millet cultivation activity in no way can compete with alternative income sources, cultivation continues to be given highest priority and wages are only considered a relevant supplement, if the harvest is insufficient.

The investigations performed so far in Bidi seem to support the statement that the agricultural system is sustainable in the sense that:1) it does not inflict irreversible damage on the natural resources, 2) there is no lack of land for cultivation and 3) the food production level is mainly dependent on the intensity of manuring (if the rain is sufficient), and manure is not considered in short supply if the man-power necessary for collection and distribution is available.

A quantitative verification, however, will demand detailed analysis of the following items:

1) Long- and short-term development in land use in order to verify whether there is a permanent shift in the location of the fields from dunes to pediplain, or whether the cultivation intensity is increasing and the fallow losing its importance.

2) Correlation between soil erosion and land use on different soil types, especially with emphasis on assessment of the irreversibility of the changes listed in a).

3) Correlation between cultivation intensity and soil-nutrient changes.

4) The relative importance of manuring for the total nutrient budget of the millet production. A proper empirical documentation of these important points is needed before firm conclusions concerning sustainability can be drawn.

Literature cited

Krings, T.F. 1980. Kulturgeographischer Wandel in der Kontaktzone von Nomaden und Bauern im Sahel von Obervolta. — Hamburger Geographischen Studien, 36.

Milleville, P. 1980. Etude d'un systeme de production agro-pastoral sahelien de Haute-Volta. — ORSTOM, Ouagadougou. Mimeo graphed paper.

Peretti, M. 1976. Projet mise en place de l'O.R.D. du Sahel. Situation actuelle de l'ORD. — Ministère du Développement Rural, Ouagadougou.

Pastoral Associations and Natural Resource Management

Poul A. Sihm
PAS Consulting, Denmark

Introduction

The subject of dryland management and pastoral and agropastoral production systems is complex because it has at the same time interacting political, ethnic, cultural and ecological aspects. The early projects of the World Bank were unsuccessful based, as they were, on the prevalent assumption of the early 1970s that development was a question of a transfer of western technology. Today much have been learned from our own mistakes, from research scientists, anthropologists and from Non-Governmental Organisations, who have brought to light the linkages between the above-mentioned aspects. An increasing exchange of experience, a multitude of studies and monitoring of field projects have created a wealth of knowledge. The Sahel Agriculture Division of the World Bank has put together a coherent strategy for natural resource management in the Sahel. A strategy which includes pastoral- and agro-pastoral resources, decentralization of decision-making to producer organizations, provision of basic extension services, land tenure reform and creation of an enabling environment including legislation, marketing, pricing and export policies.

The Pastoral Association

My definition of a pastoral association (PA) is a legally recognized voluntary association of pastoral or agro-pastoral livestock owners, who use the same free access area for grazing and frequent the same watering points. The purpose is better management of their natural resources (water, grazing, vegetation and livestock), procurement of inputs and services and the selling of their products.

It is important to state that a PA is not a project creation. An alien concept of pastoralism is doomed. I have from the outset visualized PA development only based on the existing traditional production systems, including traditional decision-making and grazing and water rights. In practice this means that an association will be made up of a varying but limited number of pastoralists bound by traditional ties of doing things together. Present numbers vary from 30–50 families in Niger to 1,000 families in Mauritania.

The objective of a PA is the security of the community and its livestock and to promote wealth of the group by a better management of its natural resources, including livestock. The PA acts as a procurer of animal production inputs, a provider of first aid in animal health and of credit to members in need. The PA can enter into partnership with donors for well rehabilitation or construction and for restoration of the environment.

The pastoralist desires food and forage security, human health services, control over his grazing but free movement in the case of drought, access to veterinary drugs, adult literacy training, schools for his children, and a reasonable price relationship between his product and the inputs he needs for the maintenance of his family and herd. This perception is only partly shared by the government bureaucrat, who desires complete control over the unruly and ignorant pastoralists, or the development bureaucrat who sees environmental degradation as a result of over-stocking, as well as disease and cyclical drought.

The tension between pastoralists and agriculturalists on the one hand and the government on the other would never have occurred if the respective post-independence governments were not guilty of serious negligence, exploitation and even persecution of the pastoralists. The failed, self-seeking development activities perpetrated over the years with the help of donors fully aligned with government policies, or lack of the same, have had tension creating effects. The development agencies refused for many years to deal in public with questions of corruption and extortion which preyed upon pastoral communities and which produced most of the private savings of local and central government officials.

In our daily dealings with the pastoralists it is well to remember that they have no reason to take us at face value and to believe in our motivation and sincerity. They will try to get the best of whichever deal you offer them and may or may not keep the promises we make them give us. They will deal honestly with you only if they feel there is a chance to achieve once more the priority right to control the use of their habitual grazing and water points. Not only must they believe you, but they must believe that there is a chance that the government will deliver on the deal you propose, and on this subject they are often better informed than you are.

Establishment of a Pastoral Association

Before entering into a description of the establishment of a PA it is important to state that no standard approach exists. Even within the same country or region differences in culture, politics, administration and natural resources require a tailor-made approach.

The establishment of an association takes time. It is usually carried out by a team consisting of an anthropologist, an animal production or health specialist with intimate knowledge of the pastoral environment, and somebody who can make a simple map of the areas claimed by the associations.

First it is necessary to gain the confidence of the group. Then the group takes its time before making up its mind. Experience shows that it often takes a whole year to get to first phase. The first visits try to establish the priorities and the constraints of the group and through a long dialogue to help the members establish a hierarchy of actions for the removal of the constraints. This initial dialogue is followed by a series of internal meetings, enough of them to create a common position for the establishment of a PA. Occasional visits to monitor this important part of the process and to provide additional information as required may allow one to judge when the time is ripe for further action. In the meantime a very positive action would in many cases be a series of literacy courses. There is no point in forcing the initial stages of the establishment of the PA. If there is no unity of purpose among the group we shall be wasting our time, and theirs.

There are a number of prerequisites for the successful establishment of a PA. They are in order of priority: legally recognized organization, allocation of resources, control over water and grazing and food and health security for the community.

Until reliable data become available I will argue the points in my usual subjective manner as follows: a) pastoralists are organized but not legally recognized and in this way they will continue to be oppressed and exploited. But the day the chairman of a PA cooperative, representing 50,000 pastoralists walks into the governor's office then his cooperation will become recognized, b) when the pastoralist knows, that he has legally enforceable priority rights to the grazing and water, which he traditionally exploits, and that he can refuse access or charge herds which have no traditional rights, then he will take an active interest in the management of natural resources, c) build a clinic and make access to payable medication easier, train midwifes or delivery helpers and see priority attached to human health, d) make forage or fodder available when a drought occurs and study what they are prepared to pay to save the breeding herd and e) visit the traditional heads of pastoral groups and see leaders wrestle with the problems of a better quality of life for their group including concern for a degraded environment.

The highest priority in this workshop, the environment, comes last when the pastoralists are concerned. Whereas our priority is the reduction of grazing pressure and resource management, their priority is security and the power to manage their traditional grazing areas. It will stay that way until the pastoralists achieve political influence and there is an allocation of resources.

In my opinion the pastoralists are well aware of the degradation of the pastoral grazing resource brought about by increasing population and restrictions on movement across borders. They are also prepared to do something about it. The first two rules to be enacted by Mauritanian PAs were heavy fines for grass fires and the cutting of trees for goat forage.

A drastic change in the ownership of herds has taken place over the last ten years because of the terms of trade which have been turned against the pastoralists. On every occasion when a drought is serious enough to force the pastoralists to sell off their herds, people whose incomes are not dependent on the vagaries of the climate, buy up animals and create herds for themselves. Apart from some very successful traders the majority of these investors are government officials,from the smallest to the biggest. They keep the animals alive and watered during the remainder of the drought. When the drought gives way to the recovery of the range, they run these herds in competition with the surviving traditional herds. These new owners are often well-connected politically and are therefore able to move with impunity their newly acquired herd round as they please. For them the access to "free" grazing during the rainy season is essential and they are willing to use all their influence to achieve it. To make matters worse, the crop producing farmer has changed his habit of not keeping cattle, or have them herded by a pastoralists. He is now teaching his son how to manage cattle. Not surprisingly they are all opposed to the organization of the pastoralists. The existence of PAs with authority over their traditional grazing would prevent outsiders from enjoying a "free lunch" during the rainy season, or at least make it payable.

In Mauritania we see this competition expressed in the amount of political obstacles put in the way of the traditional pastoralists who want to create a national union. The problems are created by a powerful lobby of "modern livestock producers", *i.e.*, the above mentioed investors, who have already created their union which the former refuse to become members of. The pastoralists know the problem and have experienced the futility of trying to defend their traditional rights to grazing and water in front of a herd belonging to a high government official. The animals are herded by a hired herdsman who often is the former owner and maybe even a relative of the pastoralists. In the absence of support from local government their endeavour becomes even more futile.

Under such circumstances it is neither reasonable to expect participation of user groups, nor is it feasible that natural resources management shall be achieved. If land, grazing and water constitute free access-areas no pastoralists will have any interest in investing labour and money to improve it. If the free roaming of investors' herds jeopardizes any improvement which may be achieved by a PA, if the price relation of production-inputs to market prices for livestock are turned against the pastoralists, and if prices are further exacerbated by the dumping of surplus production by the common market and others — how can we expect the pastoralists to adopt any other strategy than hoarding of cattle to ensure survival when the next drought comes round?

The dialogue between the pastoralists and the assisting team should convey to the groups that there is a long row to hoe before they can achieve their highest ambition, land tenure. The PAs must learn to master many things, before they can even start to make progress, most of all to act and to manage resources jointly, including money. This is where the joint management of a village pharmacy, or a small project can be a valuable experience. As an example, in Mauritania the joint action falls into three steps; the village livestock pharmacy, a small project designed by the association, and the repair and construction of a small number of wells. These activities are financed as joint ventures with increasing financial participation by the PA. The governments' participation is in the form of subsidies which generate internal savings. The principle which is explained to the pastoralists is: "You participate directly with 20% (10% cash down and 10% in kind), the government will finance the remainder on the condition that you save what corresponds to the 80%, or less depending on the the total cost of the project, and deposit it in your revolving fund over a period of years. By following this, one day you will become financially independent. If on the other hand you squander the assistance which you receive, then we shall no longer deal with you."

Pastoral associations and natural resources management

Is there indeed a potential for PAs to play a role in natural resources management? It is too early to tell, but we have learnt some facts which point in a positive direction. Anyway, how else would you organize the management of the dryland's natural resources except by co-operating with the people who exploit them? I believe it can work, but it demands long and arduous preparation, allocation of priority rights and responsibility for management of their traditional resource base, changes in national policy, pastoral organization, legal underpinning, transparent administration, attractive market and financial assistance. It will be for each country to make up its mind and, frankly, until a country is willing to take pastoral development and natural resources management seriously, I would not spend aid money on pastoral development except for demonstration purposes.

The rehabilitation of degraded range may well cost anywhere from US$ 150–600 per hectare. An overwhelming financial challenge for capital-rich but cash-poor pastoralists. Cost is a real problem where rehabilitation is concerned. Technical assistance and extension are required. But as development costs go they do not amount to the millions which the regional development banks require to function efficiently. Long term action and low cost development with slow disbursement is not an ideal menu for the regional development banks. The cost of preparing a 50 million dollar project for presentation to the World Bank's board is the same when a project is costed at 15 million dollars. But it could be just the thing for bilateral donors with a political commitment to the Sahel.

In my opinion a division of labour is called for. The bilateral donors should assume responsibility for technical assistance, research and micro-economic development. The role of the regional banks is to structurally adjust, *i.e.,* induce political, financial and administrative reforms in order to create the enabling environment in which the bilateral donors can achieve sustainable development.

Sustained development in pastoral and agro-pastoral production systems must direct itself towards mixed farming without fences, but with full participation from the pastoralists. In this field a lot more monitoring of indigenous mixed farming systems is required. The former colonial powers have spent much time and energy on the Sahel agricultural research, but there is a need for new thinking.

To sum up my conclusions and recommendations: intensified monitoring of development involving user groups in pastoral and agro-pastoral projects, division of labour between development banks and bilateral donors, long term but low cost development on the basis of structural adjustment, new thinking in the Sahel agricultural research, more debate, and closer coordination of Nordic natural resource management projects.

The UNCED Process and the Sahel

Sofus Christiansen
University of Copenhagen, Denmark

The UN and the Sahel

The interest of the United Nations system in problems of the Sudano-Sahelian region has already manifested itself in several cases.

Following the catastrophic development during and after the severe Sahelian droughts 1968–1973, the UN General Assembly 1974 recommended measures to be taken to relieve and address the problems. This led after some time to the approval of two mandates: one was to initiate drought relief and prevention of the disastrous effects of drought, and the other was to combat desertification. The two mandates were bestowed on UNSO, the United Nations Sudano-Sahelian Office and partly also on UNEP, the United Nations Environmental Programme. An interesting aspect of the mandate on desertification was that it was coined on a holistic perception of both the situation and the ways to solve its inherent problems, contrary to the prevailing, rather simplistic, views such as the narrow view on the protection of whales.

In the case of the Sahelian programme, the initiative was based on the catastrophe in the poorest of the developing countries with a death-toll exceeding 50,000, the background for the UNCED, United Nations Conference on Environment and Development, is quite a different one.

During recent years an increasing amount of public interest in the rich countries has attached itself to the problems of environmental degradation. Early expressions for this were reports such as *Limits to Growth* and the later *Blueprint for Survival*. Although a certain "overkill" was undeniable in their thinking, *Limits to Growth* presented scenarios in which a dozen factors were identified, all leading to human extinction, they successfully drew attention to the connection between present type of economic development and environmental deterioration.

In UN context the problems were treated by a commission led by the Norwegian politician Gro Harlem Brundtland who in 1988 presented its findings in *Our Common Future*. The report caused enthusiasm as well as a considerable amount of scepticism. It was difficult to grasp the concept of sustainable development, especially when attached to social and economic systems, and it was even more difficult to see how problems so convincingly depicted in the report could be solved at all.

The UNCED conference

The UNCED conference is an initiative taken by the UN to suggest appropiate solutions to avert an ecological catastrophe, which is threatening to destroy the current civilizations on the globe, if left unattended.

The instrument for this initiation will be a set of international agreements that will seek to change behaviour in such a way that development is made sustainable, *i.e.,* the natural resource base will remain undeteriorated. In fact two sets of documents should be produced: 1) *The Earth Charter*, delineating principles for international cooperation in conserving the environment and 2) *Agenda 21*, a paper in which the main ecological problems shall be diagnosed and interventions to address them outlined. The agenda should thus be seen as an international plan of action for sustainable development.

It might be worthwhile to look a little further into this international script for future international behaviour, as it will be necessary to change idealistic thinking into practical actions. Some of the topics for the negotiations are: 1) protection of the atmosphere, 2) protection of soil resources, 3) protection of biological diversity, 4) protection of the oceans, 5) protection of (fresh)water resources and 6) responsible management of wastes. Apart from these, other topics will be negotiated, mainly on questions of legislation and funding.

Some of these topics are traditional professional categories, soil, water, air, while others are modern, diversity, management of wastes. However, the heading of soil resources also covers problems of the worlds forests and desertification. The topics have been seen as "mainly expressing views of the rich nations" by the developing countries, and many of them may be seen in this light. Those that from the very start aimed at gaining international recognition of specific types of solutions, as for instance the carbon dioxide question, may especially raise distrust. These concernes have been raised in the frequent head question. "Is the idea just to keep the developing countries from getting the same technical advantages as the rich countries?" as it has been expressed.

With regard to the questions of addressing desertification and drought there seems to be no such bias, but different problems are at play as has been clearly demonstrated during the UNCED process.

Nordic involvement

Denmark has agreed to assist in the preparations for the UNCED conference to be held in Rio in June, 1992, regarding a few topics: (fresh)water resources, preservation of forests and desertification and drought.

The manpower attributed to these tasks have been rather limited, as for desertification only three persons have been allocated on a very limited time-basis. This has in some ways been compensated for by relying quite extensively on Nordic cooperation. Fortunately, points of view have always been quite similar among the Nordic nations as well as the outcome of two Nordic meetings and Nordic cooperation at a meeting held in Geneva has been encouraging. In fact, Nordic views have exerted a visible impact on the forming of general standpoints as, *e.g.* those of the UNEP.

At the present workshop of Danish researchers studying Sahelian topics, it may be worth noting that the points of views expressed on several occasions mainly have been such as emerged from experience gained from research. An example may be referred to. At a November 1991 meeting in Geneva, the UNEP presented two draft documents: 1) *Status of Desertification* and 2) *Implementation of the UN Plan of Action to Combat Desertification* (UNPACD).

In the status-document two very obvious issues clearly were observable; the lack of a proper definition of desertification and, partly consequential, the enormous problems in arriving at just reasonably accurate assessments of the extent of desertification.

The UNPACD-document revealed great difficulties in establishing priorities in the combat against desertification. It also gave the impression that the UNEP approaches to the concerned people were still the traditional "top-down" approaches whereby the locals are very sparringly involved in the work to save themselves. In the background of the Nordic preparatory meetings, especially the statings of the status document were criticized. The definition of desertification was challenged and the estimates of areas affected by desertification were found untenable.

Instead, an improved definition was proposed or, alternatively, it was proposed to disuse the term completely. Desertification was suggested to be defined as: 1) land degradation in arid, semi-arid and dry sub-humid areas mainly resulting from adverse human impact, 2) land degradation in this concept includes soil and local water resources, land surface and vegetation or crops and 3) degradation implies reduction of resource potential by one or a combination of processes acting on the land. These processes include water erosion, wind erosion and sedimentation by these agents, long term reduction in the amount or diversity of natural vegetation and salinization and sodication.

This type of definition was, without much resistance, accepted by the UNEP, which may mean that the advancing desert frontier finally disappears from most publications. Clearly, the estimates of desertification may in the future be based on methods for assessment, that at least is based on a reasonable definition: the first requisite to use any method for estimating. Similar definitory problems apply to dry lands, arid, semi-arid and dry sub-humid areas.

Regarding the last definition(s), as is widely appreciated, quite a few definitions are available. Most of them may need refinement, but it is at least evident, that for comparisons identical definitions must be used. This is not the case in the UNEP's status report, so conclusions on whether desertification is expanding or not are extremely dubious. Nevertheless, the UNEP in its first report gave rather precise measures of areas in different degrees of desertification.

The UNPACD-document revealed difficulties in chosing priorities for interventions to stop the spread and worsening of desertification. The priorities set were to rehabilitate degraded lands in the order of: irrigated lands, rainfed crop lands and rangelands, respecting at the same time the seriousness of degradation.

This was seriously doubted to be rational, since the availability of funding (presently about one billion US$ per year) by no chance could cover such a vast undertaking before year 2000. Instead, it was suggested that the least affected, but desert-prone areas were given highest priority within all categories of use.

Research and monitoring

The conference also discussed the necessity of monitoring and further studying of the processes of the on-going degradation, including such research in socio-economics that may be seen necessary to grasp the roots of the poverty and the behaviour of the population. Clearly, the ideas of the UNEP are that all monitoring be under its auspices and applying its methodology. Many representatives were apt to see monitoring being best cared for on a national basis, with international support and voluntary cooperation. The following may possibly be seen as those programme points likely to be the final results of the discussions: 1) strengthening of the knowledge base and developing information and monitoring systems both on physical/biological topics and on economic and social ones, 2) combating land degradation and intensifying afforestation and reforestation activities, 3) developing and strengthening integrated programmes for the eradication of poverty and promotion of alternative livelihood systems in areas prone to desertification, 4) developing and integrating comprehensive anti-desertification programmes into National Development Plans and National Environ-mental Action Plans and 5) developing comprehensive drought preparedness and drought relief schemes for drought-prone areas and design programmes to cope with environmental refugees.

In this context it is probably mostly the topics mentioned under 1) that may attract attention. Within that area the most prominent activities are likely to be:
1) a permanent global monitoring system to be established within EARTHWATCH (UNEP's system),
2) a global assessment of status and rates of desertification to be finalised

before year 2000,

3) local and national capacities be established within research and monitoring, *etc.*,

4) UNEP to guide studies on social and economic consequences of desertification and interactions between climate, drought and desertification, to arrive at better definitions, concepts and assessment methodology,

5) national governments should support assessments of soil degradation, status of vegetation cover, and supply data for the desertification database components of EARTHWATCH. Meteorological and hydrological networks shall be strengthened and

6) cooperation with CILLS, Comite Inter-Estats de Butte Contre la Secheresse au Sahel, IGADD, Inter-Governemental Agency on Drought and Desertification, SADDC, Southern Africa Development Coordinating Conference and OSS, Observatoire du Sahara et du Sahel, should be promoted.

Everywhere existing inventories and information systems should be integrated and used. As appears from the above, Danish researchers may well be seen as energetic suppliers to the programme.

Beyond the Agricultural Crisis in the Sahel

Moussa Seck, Phillipe Engelhard and *Taoufik B. Abdallah*

Environmental Development Action in the Third World, Senegal

Introduction

The following account is somewhat particular in that it is a review of a concrete experience which, at the same time, is the outcome of a reflection. In itself this is not very original, except that this reflection is of a prospective nature.

We belong to a team whose main mission is prospective. Rather than predicting the future, it is often tempting to create it or to facilitate its creation.

The rural crisis

This prospective action originates from a fact and an opportunity. In Senegal, as in other parts of Africa, the rural areas are in a state of *crisis*. This crisis has two aspects: 1) The economic aspect, the most representative cases of which, the farmers' gross income and productivity seem to have stagnated and 2) The population aspect, in 20 years the population has doubled.

These facts are not a crisis in themselves, but are indicative of poverty and lack of development. There is not famine. The farmers, however, are having difficulties sending their children to school and even just taking care of them. The widespread poverty, artificial urban expansion and the incredible number of young people on the continent have caused questions to be raised about the social and cultural structures.

Just to clarify, when the word crisis is used, it is not to imply a catastrophy at least in the common use of the word. There is simply a rupture. This rupture is also a source for new opportunities and, perhaps, a renewal of the rural areas. Before talking further about this possible renewal, the so-called social rupture will be summarized. On the one hand, as a result of the high populations growth rate, the continuity of cultural generations is being upset (50% of the rural population is less than 18 years old). The youths still adhere formally to the social values of their parents, but do not belive deeply in them. On the other hand, education and the attraction of the cities have serious consequences. Firstly, the youth migrates to the cities. Today they are the ones who redistribute their profit for the benefit of the older generation and the parents in the villages. The sense of redistribution and power has been reversed. Secondly, adults remaining in the country find themselves more and more demobilised. Shocking as it may be, reality is daily revealing that villagers are believed to be less and less inclined to make an effort. Formerly, farmers used to stay in the fields from dawn to dusk, at least during the cropping season. The time spent in the fields now has been drastically reduced. Economically, psychologically and socially farmers are being demobilized. This is the true crisis. It is tempting to ascribe this crisis to the lowered incomes and especially less fertile soils.

But the problem goes deeper. It should, however, be mentioned that considerable enthusiasm is being shown in the associative movement in Senegal as there is on the African continent as a whole. Its extent should not be underestimated. A great part of this movement is stimulated by aid that needs to be channelled and that the movement for the time being consists more of a system of redistribution than a system of production. The movement is without doubt stimulating the will to change, but it can not alone create the change.

Deep in our minds we do not really believe in community accumulation strategies. We believe that it is on a closer family level that new strategies have some chance of application. The rural population can be activated. This activation, however, often demands a push from the outside, but different from what is currently prevailing. The push should consist of the capacity to mobilize new social energies.

Possible outcomes

Perhaps this crisis carries, if not the components of its own resolution, components that can lead to possible outcomes. In our point of view this outcome is at the intersection of three key ideas.

The first key idea consist of five fields: a) the employment field where less than 10% of the urban youths (35% of the total population), who are often of rural origin, are employed by the modern sector, b) the field of potential freedom to innovate and produce in the rural zones, c) the field of liberty, now elders and adults are less inclined than before to imprison youths in rigid social and productive systems, d) the field of production, yields are low, potential demand of agricultural products are high and production systems are very often under-used due to insufficient demographic pressure and e) the field of considerable needs, inside as well as outside the country. One thing is certain, however, distribution is inadequate and therefore the latent market is unsatisfied.

The second key idea is that few synergies of production have thus far been attempted. Development very often leads to a juxtaposition of disconnected projects. Yet, true development is obviously the art of linking very small pieces together, therefore it deals with engaging various processes: agricultural and artisinal productions of synergies and commercial, spatial and informational relations of synergies. Thus, it deals with the creation and distribution of income which are a source of a more developed solvent demand. In short, self-sustaining supply and demand is what should be stimulated, especially those of an agricultural productive nature; other motors of this kind can be imagined in urban zones.

The third key idea is about self-sufficiency. Self-sufficiency being an end itself, is perhaps the worst way to address the problem of agricultural development. In our point of view, the problem is not how to feed everybody, but how to make good farmers, who from their production can earn enough to cover their own consumption as well as have a surplus. If we have good farmers who earn their living correctly, we will furthermore increase self-sufficiency, or at least food security, as a result. Self-sufficiency as a goal in itself is the starting point of a development strategy. Traditionally, farmers are too often inactivated by the aforementioned crisis. Contrary to what is often thought, classic productions systems exemplify the inventor's greatest imagination. Today, however, they are outdated. Rural youth as well as city youths can be viewed as a new work-force which has to educate itself in several synergetic directions.

An outline of an outcome

The agricultural projects that we follow present three major characteristics.

Technical characteristics. — These production systems are designed from traditional models which sought the most rational and efficient expression from a technical and social point of view. Schematically, it is a question of using area and time optimally, which brings foreward the principle of crop rotation and mixed cropping.

Without going into details here except for an example of a growing season while maintaining the fundamental idea of multiplying the production area and diminishing the production time. The Daga project, located 50 km south of Dakar, Senegal, where on one hectare of land we grow fruit trees, grains, vegetables and fodder. On the arboreal production system level, five species are chosen according to their various maturing time, Sour soap, Sweet soap, Avocado, Mango and Grenada fruit. Vegetables grow from October to June in two successive seasons. From June to October cereal grains are planted alternately with millet, sorghum and corn. The advantage is that grains benefit from fertile residues not used by the vegetables they replace. A considerable advantage follows from the entire system. A doubling or even tripling of the grain yields is achieved, compared to the traditional systems (1–1.5 tons/ha can be obtained). The growing season is organized in such a way that everything takes place as if the farmer was using the traditional four hectares instead of one.

The main benefits are better use of areas, concentrated efforts and continous use. Owing to crop combinations and symbiosis among species concerned, the farmer will produce in one year, what would have taken five years with traditional production by playing with time and area. Potential yearly gains: F.CFA.[1] 4–5 mill. according to the current state of supply and demand.

Social characteristics. — The participants are young, educated urban and rural people who must take up a new way of farming; we are referring to them as the "young farmers of tomorrow". The organization is composed of the individual or nuclear family size, production and collective distribution and a sales system.

Since these systems require a relatively sophisticated organization, they call for participants who have a sufficient level of education. Education has often been observed to be a factor of migration. If the farmers have sufficient incomes, their children will have no preconceived intention of earning their livelihood in cities, often without luck. In any case, if education is conceived as an instrument of communication and not as an academic "obstacle course" there would be less migration. Furthermore, one wonders how to organize a relationship between new farmers and their older counterparts. But it can be reasonably imagined, as experience seems to confirm that adults 30–40 years of age are taken in by these new systems and older people adhere to them, at least, informally.

Procedure and interdependence characteristics. — Rather than stating abstract points of view, three examples can be shown:

1) The *primum movens* in agricultural production lies in the form of Economic Interest Groups. It seems evident, that local artisans could easily make a good deal of the needed farming equipment at reduced costs. This is an obvious way to re-use the value added and profits.

2) We support the idea that the products' ability to sell depends on their quality. We have therefore created a system of conditioning fruits and vegetables which will start producing the following year. This is yet another source of income distribution for women and children and thus a potential demand.

3) The agricultural production system is capable of producing animal fodder during the entire year. In the Toucar project 60 heads of cattle were fattened as a result of a mixed diet. We are currently setting

---

1 50 F.CFA is equal to 1FF. Editors remark

up Economic Interest Groups of young urban butchers, able to purchase good quality pasture cattle and slaughter them according to new consumer demands and rules.

Other synergies can also be imagined, some of which we are attempting with 1000 young farmers who are located in different regions of Senegal.

Conclusion

Production systems and their radiating synergies do not constitute a panacea. Their success among youths lead us to think of the following: agricultural development was often conceived in terms of separate projects designed to make improvements in the farmers life in a traditional framework. The social, cultural and economic rupture we are seeing should obviously inspire us to imagine something else: new production systems in new social areas. Senegalese farmers have seen great revolutions in the past, but today they are finishing the last of the efforts of the past. This current rupture needs to be turned into profit and creation of something new, not by breaking family ties, but rather by aiming to polarize them in another direction.

Again the self-sufficiency idea is linked to famine. Africa is not in a state of famine, except for certain areas. It is in a state of poverty. Self-sufficiency can only be something to fall back on. Above all, our ambition is to set up a solid family-farming system, which is income-generating, competent and therefore competitive on the national as well as on the international market. The emergence of such an agriculture which can only be ecologically sound is without doubt the best way to preserve the environment. It will preserve it, because it will modify it in an appropriate way.

Sustainable Development: just An)ther Pet?

Søren Leth-Nissen
Provincial Fish Culturist, Zambia

Introduction

Until now, a proper definition of the term su ;tainable development has not been made. Since the term is widely used, ʃ find it necessary to try to define the concept. The concept connotes the n 'o-functionalistic and neo-evolutionistic nomenclature and paradigm. If we critically analyze the concept within this framework, the difficulties of creating or just even talking about sustainable development become very apparent.

Sustainable development — a new catchword?

The honoured professor Martin Paldam, Aarhus University, Denmark, once said that western donor agencies, NGOs, planners and, not least, the public need so called pets in order to be able to accept the amount spent each year on aid to Third World countries. We need an easily-identified catchword to help us forget that the Western World earns huge heaps of dollars through this so called aid.

An example: in 1988 developing countries received US$ 48 billion, but they paid US$ 330 billion back. A shipment of average African goods in 1980 could be exchanged for 100 tractors. Africa got 86 tractors out of that deal. Only 86, because 14 went to pay interests and down-payment on debts. However, in 1989 the same shipment could be exchanged for 62 tractors only, of which 30% (19 tractors) went to payment of interests and down-payment on debts to foreign creditors. That left only 43 (62 minus 19) half of what was left nine years earlier. One could ask if aid should be called business instead?

The wolves (critics) are right at our necks and we have to invent still new catchwords to keep the wolves away from our romantic ideas of helping the Third World.

After the Brundtland report, the question is whether the concept of sustainable development is just another pet or if it is really possible to obtain a sustainable development?

Over the years there have been many pets. In random order: the green revolution, women, peoples participation, democracy, market oriented economy, human rights, and now the key word is sustainable development. The problem with pets, though, is that they die. They are not sustainable, just like development. Development is here seen as retrospect development and predicted development. In retrospect development, explanations are created through collection of selected factors usable for the writing of development history. Predicted development is seen as the ideal way things should turn out. In other words, ideology.

All lines of development, going back or forth, will eventually break the line of intended development at a certain point and new lines will be formulated, explained or designed. Development is not and never was, sustainable. It might be more appropriate and humble to replace the term development with the term change.

The ecosystem model and the upcoming social science

In 1935 an ecosystem model was established in accordance with Charles Darwin's (Darwin, 1859) theory of evolution. The nature was, so-to-speak, put in boxes and a flow-system of energies was illustrated. Seemingly, the model could fit in all living organisms and even demonstrate that Darwin's theory was right since the mutation process was seen as a mere adaptation of a species into another box (niche) in order to survive. This model was very illustrative and suddenly it was much easier to describe and explain nature and some of the possible functions of it.

American educational institutions, such as Columbia University and Michigan University, grabbed these models as well as Darwin's adaptation concept and started to show functional relations or linear models in different cultures and created an immense amount of both useless and useful data. The ecosystem model and the adaptation concept were rigidly transferred to the social sciences, and cultures were from then on in American anthropology described in functional equilibrium paradigmatic terms.

However, the problem with describing societies in such a way became rapidly apparent. Either by studying the history or by simple observation it is evident, that people do not behave according to the natural laws, described by physicians, biologists, anthropologists and the like. In order to make the model mimic reality they had to fixate the different variables such as inter, intra and external constraints which meant that they closed the systems and kept on operating with linear models.

The neo-functionalists could not explain why a society changed when it had just been described as being in an equilibrium and sustainable state. The neo-evolutionists tried to operate with different levels of evolution determined by the ecological and geographical conditions.

However, an important conclusion was drawn, n mely that social systems cannot be described in natural science terminology such as a linear model, since the empiric data show that societies change through time in other directions than the expected.

Ecosystem model, Club of Rome and environmentalists

Even though critical lessons were conducted by mainly French social scientists showing that societies are dynamic, the wave of describing cultures as carrying capacity linear models raved on. The Club of Rome hit the new pet-words when they published thei *Limits on Growth*. The book shocked the whole (western) World. The result was that when the environmental movements took up the same pl rases and models as the social scientists were using (the ecosystem model), the ecosystem niches were taken as real entities in nature. They started using concepts as "balance of nature," "original ecosystems," *etc.*. An original ecosystem was described as a state of nature where the nature keeps the equilibrium through self-regulating mechanisms. Further, it is defined as an entity in nature which is not exposed to human influence. If these mechanisms are disturbed, the system will loose its balance which again may cause the collapse of the entire system with extinction of the (sub)ecosystem as a consequence. The main conclusion is that man is the only factor that can destroy the balance of nature. The logical consequence is therefore that man can never be part of the balance of nature. This means, however, that man can never build a civilization that will not create environmental problems. It also means that we cannot talk of sustainable development as long as this paradigm is the ruling one. As long as we discuss within the framework of linear models, carrying capacity, sustainability, *etc.*, we define the problems as unsolvable.

The famous and respected Brundtland report uses the same concepts as just described and the very concept of the sustainable development is put forward as if this is the solution to the problems that humanity faces. If it was possible to operate with the concept it would be great, but the concept is formulated in an unconstructive way. So why not just leave it?

Towards a new model for understanding

The alternative to the use of linear models is to operate with dynamic models and theories. This implies that we work with variables that remain variables in a multidimensional open system. This means that life becomes much more difficult, complicated and unpredictable. But we need to realize that life is not what we think nature tells us it is, but what we try to make it. Ergo, a political question. If we decided to eliminate the risk of destroying the ozone layer, we could in principle agree on it and just stop the freon leaks. All it takes is a little bit of suffering, mainly in the Western World.

Sustainable development as a social scientific tool does not exist and will never come to exist until we discuss the concept more thoroughly. Instead, we could talk of the development as an ever changing process where we have to change the objectives and the means constantly in order to fulfil the formulated and nonformulated policies that we in the Western World agree on (read my lips: dynamic models). (It would be an exaggeration to state that the Third World countries have anything to say next to the Giant Seven countries).

The actual development process cannot be predicted, but we can be aware of that man, to a large extent, himself decides how the world we live in should be.

Literature cited

Darwin, C. 1859. The Origins of Species by Means of Natural Selection (OR the Preservation of Favoured Races in the Struggle for Life. This edition published 1979). — Avenel Books, New York.

The Nile, the Desert and the People — towards a Cultural Strategy for Environemental Action

Søren Skou Rasmussen
COWIconsult, Denmark

Introduction

Various strategies for environmental action in the Sahel region are discussed and visible during the planning and implementation of projects and programmes. Professionals within various disciplines formulate a strategy for the community concerned.

This paper will demonstrate that although much attention is paid to the technical solutions, and often also to the need for participation by the communities concerned, the cultural strategy towards environmental changes is often overlooked.

Anthropologists and sociologists examine the social relations and structure of the community which is the base of the cultural strategy. Often, however, their studies, recommendations, conclusions, *etc.*, have little impact on the planning and implementation of projects and programmes. This will be discussed further in the paper and possible solutions will be suggested.

The specific case examined in this paper is the UNSO project "Afforestation and Reforestation in the Northern Region", Sudan. The project addresses the village communities along the River Nile, the Atbara River and the Letti Basin which have a population of 105,000.

The main problems facing the communities are sand encroachment of villages and agricultural land. The objectives of the project are to control this sand encroachment through establishment of shelterbelts, windbreaks and land management activities with a high degree of participation from the local people.

The environment defined as the interaction between the Nile, the desert and the people

In short, the local environment of the project in question can be defined as the interaction between the Nile, the desert and the people. The Nile constitutes the natural resource, being utilized by the farmers living closest to it, to irrigate their fields, while the nomads living in the desert further away from the river use the water for their animals. For this purpose the nomads have to trespass onto the agricultural land. Farmers and nomads interrelate when farmers are exchanging grain for animals. This way of sharing (managing) the natural resources based on a geographical division of resources is a long established tradition.

Parallel to the natural resource management there are social structures of the communities involved. In this context it is not so much the social structures of the farming and nomadic communities which are important, as it is the symbols used by them which are set at stake when acting in relation to the structures.

Seen from the direction of the Nile and the farmers, which is the logical direction of the project, the desert is distant, unlimited, non-productive and wild, — it is true nature. Therefore, the people living there are strangers, "arabs", a word referring to its classical connotations of simplicity and generosity of true desert people. The farmers have no doubt that the "arabs" should be found at a stage previous to their own civilization; the "arabs," thus, are not uncivilized but rather precivilized. The "arabs" and the desert are becoming a past which at the same time, symbolizes an origin and a past left behind by the farmers. The farmers conceptualization of the desert and "Arab" is something representing a wilderness and a threat forming part of a reality to be protected against.

Environmental change and cultural strategy

The sudden emphasis on the many environmental projects and programmes initiated by the government and the donors have left the population with the impression that the whole environmental issue, and even the environment itself, with its changes and its need for protective measures have emerged from the programmes. This has often left the communities involved in these projects and programmes in an unfortunate position of not being heard or not being understood, because their concept of environment is formulated in a different way. Not only has our concept of environment been dominating and often misleading, the same can be said about the contents of the concept. This will be discussed in the following.

The relations between the farmers, or "river people," *nas al-bahr*, as they call themselves and nomads or "desert people," *nas al-khala*, have always been based on a negotiated balance in the management of the natural resources, giving the river people priority to the Nile and the desert people priority to the desert.

Environmental changes occur constantly and the communities respond according to what may be called a cultural strategy. Based on the geographical division of resources mentioned above, it is the communities which in general feel the environmental changes and formulate a strategy, while their leaders negotiate the action.

When drought occurs or when the rains are late, the nomads migrate and set up their quarters on the desert fringe close to the Nile. Through negotiations with the farmers they are given passage to the Nile to water their animals, while the farmers benefit from the interrelation by selling crop residues for the animals. In years of heavy rains in the desert, many farmers move to the desert to cultivate areas acquired through negotiations with the nomads, while the nomads benefit from selling milk and meat to farmers. This is an example of a well-established cultural strategy towards a changing environment. The existence of a strategy towards smaller annual variations does not preclude the existence of cultural strategies responding to more long-lasting and dramatic environmental changes. In many cases these more fundamental changes are the same changes which the environmental programmes address.

It is a well known fact that continuous drought in the Sudan has forced large numbers of people living in the desert to move towards the Nile on a more permanent basis. This has put a severe stress on the above-mentioned balance between river and desert people and their leaders who negotiate this balance. Local and tribal sheiks and other traditional leaders are constantly consulted by farmers who see their land and their basis of living being threatened by the newcomers and their animals. In some cases direct hostilities have broken out, in other instances the newcomers have been integrated as farm labourers and unskilled workers in villages or towns. I consider the last example of migration another cultural strategy towards a changing environment.

An interesting aspect in this context is the various traditional leaders and their negotiations concerning land rights and land distribution: do they negotiate as if they believed the former balance would be re-established, or as if this new situation is a more permanent structure? There seems to be many different perceptions reflected in different strategies.

Desertification and sand encroachment also present environmental threats to the farm communities along the Nile. In some places sand dunes up to 40 m high approach the villages and the agricultural land. Archaeological studies show that this process of moving sand has taken place for centuries. For most of the communities along the Nile of northern Sudan, however, the severeness and amounts of sand present a new problem, for which new strategies have to be invented.

For the villages two main strategies are in use. If only part of the village is exposed to sand encroachment, each household reconstructs its house on top of the old one, when buried in sand, using the old house as a foundation for the new one. The result is that in some places one may find houses reconstructed three times on top of each other. When the whole village is being engulfed by sand, another strategy is applied: the decision is taken to move the whole village to a new place.

The strategy of moving has to be further explained, as several environmental projects aim at protecting villages. The planners of these projects often have their cultural idea of a house being a valuable construction in itself and therefore something which has to be protected for generations and centuries. The communities along the Nile in northern Sudan never thought that way. For them the family and tight family relations are more important than the house itself. The house is simply a shelter for the family, which can be rebuilt anywhere at any time. One rarely hears about someone inheriting a house in the villages along the Nile. Village houses represent no value as such and are seldom more than 50 years old.

In respect of sand encroachment on agricultural land strategies are more complicated. Agricultural land is, contrary to that of the houses, valuable in itself, limited and inherited. All three characteristics have an influence on the cultural strategy when agricultural land is being threatened by sand encroachment and/or by desertification.

It becomes clear that land is valuable in itself when part of the agricultural land is threatened. Next to the land threatened by sand encroachment is land which is not affected by sand nor cultivated and therefore has a potential for cultivation. Nevertheless, the land is not available, as it is owned by one or several persons who are not willing to sell it to be cultivated by other people. The fact that land resources are limited means that movement of agricultural land is rarely a possibility or solution. The inheritability of agricultural land is also an obstacle to moving of land.

Only if land having these characteristics is available will the agricultural area of the village be moved. The available land is most often found on the desert fringe, which presents another problem as nomads often settle there owing to the same sand encroachment and desertification. Therefore, protection rather than moving of the agricultural land becomes the most widespread cultural strategy, in response to the threats by sand encroachment and desertification. Another common cultural strategy is migration of farmers and/or farmers finding another way of living in the village.

In other words we are dealing with two different principles of cultural strategy towards environmental change: the first being movement, the second being protection. Several other examples of environmental change and cultural strategies could be given, but for this purpose the establishment of certain principles of logical parameters is more interesting when focus is on environmental change and project strategies in a defined community.

Implications of project strategy *versus* cultural strategy

Without having discussing the broader issue of people's participation, it is clear in the case of UNSO's Afforestation and Reforestation Project, as in many other projects, that focus all the way from the planning phase to the implementation phase has been more on participation than on the people. We tend to think that just the participation of the people in the work means that the project is going to succeed. I will not disqualify the concept of participation of the local people, but point out the danger of a perception of people's work relations as more important than their social relations when it comes to implementation of the project. The participation of the local people will then become means to provide cheap labour, while their social relations and structures are kept outside. I will later make suggestions for how the people involved, especially anthropologists who are part of this crucial mistake, may rectify it.

The *Afforestation and Reforestation Project* presently work hard to formulate parameters for participation of local people. There is no longer doubt that participation of local people is deeply rooted in local social structures and relations. Social insecurity and stress *versus* social affluence is one parameter affecting the social relations and thereby the ability of the participation of the local people. Can the project expect participation of local people in establishing a communal shelterbelt, if and when villagers have difficulties with their physical and social survival? Can the project expect participation of the local people, including communal work by the villagers, if the villagers in a relatively affluent society are used to pay labourers for these communal services and works, while their social relations are at stake at other levels than communal work (trade, weddings, funerals, *etc.*).

The project objectives do not distinguish between environmental threats to houses and agricultural land, and therefore the same technical solutions are sought for, whether the project protect villages or agricultural land. The reason for this mistake is that the project in its planning phase never assessed the cultural strategy of communities towards environmental change. In some villages the project risks establishing shelterbelts to protect houses, where nobody lives any longer because the project recommended protection, while the village recommended moving. Understandably, participation in these villages is rather low.

However, participation is high in other cases. In some villages on the desert fringe, where shelterbelts are established with the objective of protecting village houses, interest and participation have been high. One explanation may be found in the character of these settlements. The people living there are nomads who have settled only recently. They have no land, but work as farm labourers on nearby agricultural land. They have serious disputes, some of them still on-going, concerning their rights in the negotiated balance between the desert people, which they used to belong to and the river people, which they aspire to become. For these, often economically non-privileged villages, the shelterbelt project is both a physical protection against sand encroachment of an unfortunate place they were given, or for which they have fought and a symbolic structure which is part of securing the existence of the village. This is surprising, because these people were nomads only recently with no cultural idea or strategy of protecting a place for permanent settlement, but logical when considered in relation to their changed environment.

Because the strategy of protection is also the project strategy, a most successful interaction between the project and the communities of recently settled nomads has been established.

In general, communities in the project area have had little previous experience in protecting their land. Some protective shelterbelts were established in the 1960s, but they are few and not very significant. The project has thus contributed with new technical solutions and strategies, such as shelterbelts, windbreaks and sand dune fixation. The communities are eager to participate in protecting their agricultural land, but farming communities often lack the long term perspective needed in forestry to wait for the outputs and benefits, and therefore many villages are looking for occupation outside the agricultural sector in the village.

The widespread search for jobs outside the agricultural sector leads to a situation, where agricultural land and protection are no longer given high priority. This does not mean that protection of the land looses importance. But the project has experienced difficulties with its otherwise successful application of the participation of the local people and involvement of the communities in the protection of agricultural land in villages, where agriculture no longer provides the basis of living for the majority of the people. In order to avoid situations where the project concept stresses the need for participation of local people, while people prefer migration as their cultural strategy towards a changing environment, there is a need for the project to be more careful in the selection of project villages.

For the communities on the desert fringe, which have no agricultural land, the establishment of shelterbelts gives opportunities for cultivation. This is in accordance with project objectives of increasing the cultivated area. The banks of the irrigation canals and the land close to the shelterbelt and village nursery can be used for cultivation, but for various reasons the agricultural potential is not always exploited. Thus, discussions in a village about establishment of a land tenure system on this land concluded that it was better for the time being not to cultivate the land. The villagers argued that the land was scarce and cultivation would create social and economic inequality.

All the above-mentioned implications of project implementation caused by a difference between cultural strategies, stress the need to find way of combining various cultural strategies in project planning and implementation. As the focus of this paper is on cultures rather than on technical aspects, I will discuss in the following section what the anthropologist might do in order to make ends meet.

The anthropologist, project planning and implementation

The risk of confusing social relations and work relations has been mentioned as an example of a problem, which should make the anthropologist reconsider his role in project planning and implementation.

During the project planning the anthropologist is normally involved in a sort of baseline study. This is often conducted according to well-known anthropological theories and methods, often focusing on social relations. Usually the anthropologist surveys the project area through a selection of households, community leaders, *etc.*, Although it is part of the project planning, the baseline study is often considered a separate exercise especially by the anthropologist himself, who tends to believe that his independence from the planning of more technical issues of the project is an advantage. I consider this a misconception and the present paper should illustrate the need for a very close link between the technical project strategy and its professional planners as well as the cultural strategy and the anthropologist.

In the case of "The Afforestation and Reforestation Project", the anthropologist has collected and analyzed a considerable amount of basic socio-economic information both from individual households and from the community in general. But where is the link between the comprehensive baseline study and the planning of the project? From the beginning, the planners were more interested in implementing a project by use of the participation of the local people than understanding the social relations and structures as surveyed by the anthropologist, who considered the survey important for project planning and later implementation. The lack of mutual exchange of findings and ideas made the planners and the anthropologist confuse the concepts of social relations and work relations. They thought that they were planning and implementing a project in correspondence with a prevailing strategy for the community as a whole and not, as later realised, a project only responding to the strategy used by a small segment of the community.

I am not blaming the technical planners, as is often done, though with a limited impact. Instead, I suggest that the anthropologist should change his view and reconsider the focus on the baseline study.

In case of "The Afforestation and Reforestation Project", a study of the various cultural strategies towards a changing environment combined with a thorough assessment of the relevance of project ideas and strategies would have been a more relevant approach than the one applied.

Traditional baseline surveys including investigation of the socio-economic status of a large number of households, instead of the assessment of the more or less fictive idea of the people's willingness to participate, are often just responding to the planners' ideas of what a socio-economic baseline study should be. It is the responsibility of the anthropologist to demonstrate that it could be different. Often the basic socio-economic data are already collected and analyzed by others when the anthropologist starts working. The collected data should only be considered an important background for the more important study of concepts, ideas and strategies used by the community.

The impact, status and nature of the work of an anthropologist changes dramatically from a planning phase to the implementation phase of a project. From thinking that he is on safe ground during planning and baseline study, the anthropologist often finds himself in a much more complicated situation during the implementation phase. Funding agencies and others often advise him to continue the household surveys along the same lines as in the planning phase of the project. The multi-function duties and responsibilities on the battleground itself, however, rarely leave much time for the anthropologist to continue in the "traditional baseline way." Further, he often realises that the baseline study does not facilitate his work as was the intention.

The experience and findings presented in this paper clearly point to the need for a new approach and position of the anthropologist involved in project planning and implementation. A new approach to make ends meet, the community, the anthropologist and the project.

The Role of Nodulating Bacteria and Mycorrhizal Fungi for Establishment and Growth of Trees in Ethiopia and Somalia

Anders Michelsen
University of Copenhagen, Denmark

Currently a group at the University of Copenhagen is involved in studies of the ecological effect of tree planting in Ethiopia. One part of these studies is to attempt to elucidate the soil-plant interrelations influencing the nutrient uptake of plants, below-ground processes in which nodulating bacteria and mycorrhizal fungi play an important role. The present paper is an attempt to summarize this on-going research.

The role of nodulating bacteria in enabling crop and tree legumes to fix nitrogen from the air is well known, and much research in the Third World is focussed on maximizing the benefits from this symbiosis. Less well known symbiotic organisms may be vesicular arbuscular mycorrhizal fungi. They colonize the root systems and the soil rhizosphere of the vast majority of plants in tropical and sub-tropical forest ecosystems and facilitate the uptake of potential growth-limiting nutrient elements, *e.g.*, phosphorus, by adding the fungal mycelium to the functional root surface over which nutrients are taken up from the soil solution. Furthermore, mycorrhizal fungi are capable of increasing the plant's resistance to drought and root pathogens (Michelsen 1990, Michelsen and Rosendahl 1990).

Observations, *e.g.*, from Somalia (Michelsen and Rosendahl 1989), suggest that mycorrhizal fungi are sparse in areas with degraded vegetation. This is not surprising as fungus and host plant are mutually dependent; the fungus receives sugars from the plant. In such areas the lack of mycorrhizal fungi may severely affect the survival of young trees after planting. Development of mycorrhizal colonization in newly planted seedlings is delayed because of the low level of fungal diaspores in the soil, and the seedlings will thus suffer more from the severe drought and nutrient stress to which they are often subjected at the time of planting. In one site in Somalia all the seedlings planted in a species trial died before the end of the dry season. This could be due to lack of mycorrhizal fungi in the tree nursery soil in which the various seedlings were raised (Michelsen in press). Furthermore, none of the acacias from this Somalian nursery formed root nodules with bacteria, although all species are known to have the potential to do so. A similar lack of mycorrhizal fungi was found in tree seedlings in one of four investigated Ethiopian nurseries (Michelsen in press).

In order to reveal the possibility of improving the growth and survival of seedlings by optimizing the conditions for mycorrhizal fungi and nodulating bacteria in tropical tree nurseries, several nursery and field experiments have been established in various parts of Ethiopia. The experiments involve application of isolates of mycorrhizal fungi as well as attempts to use more simple, low cost inoculation techniques feasible on the local level, as the addition to the nursery soil of chopped roots from plants growing in the vicinity of the nursery.

Most of these experiments (Michelsen submitted) have shown that it pays off to add mycorrhizal fungi or chopped roots to the normal (non-sterile) nursery soil: the growth, measured as shoot dry weight, of *Acacia abyssinica*, *A. nilotica* and *A. sieberiana* in one nursery increased with 15–45% after addition of the various fungi and roots compared to seedlings growing in the standard soil. Moreover, the survival of *A. abyssinica* seedlings after planting into the field was increased. The nitrogen fixation of the seedlings inoculated with roots or mycorrhizal fungi was enhanced in some cases too, probably as a result of enhanced nutrient uptake, which again facilitates the nodulation process and nitrogen fixation by the rhizobial bacteria. The experiences from the trials will be employed when compiling suggestions for appropriate management practices aiming at maintaining optimal populations of mycorrhizal fungi and nodulating bacteria in tropical tree nurseries and plantation sites.

Acknowledgements

This study was supported by the Danish Council for Development Research, grant no. Dan.8/481.

Literature cited

Michelsen, A. 1990. Mycorrhizasvampes betydning ved træplantning i Nordøstafrika. — U-landbrug 2(3): 10–13.

Michelsen, A. in press. Mycorrhiza and root nodulation in tree seedlings from five nurseries in Ethiopia and Somalia. — Forest Ecology and Management.

Michelsen, A. submitted. Improved growth of *Acacia nilotica* by the VA mycorrhizal fungus *Glomus intraradices* under non-sterile nursery conditions in Ethiopia. — Proceedings, Third European Symposium on Mycorrhizas, Sheffield, U.K.

Michelsen, A. and Rosendahl, S. 1989. Propagule density of VA-mycorrhizal fungi in semi-arid bushland in Somalia. — Agriculture, Ecosystems and Environment 29: 295–301.

Michelsen, A. and Rosendahl, S. 1990. The effect of VA mycorrhiza, phosphorus and drought stress on the growth of *Acacia nilotica* and *Leucaena leucocephala* seedlings. — Plant and Soil 124: 7–13.

Aspects of Nutrient Cycling in Three Plantations and a Natural Forest in Ethiopia

Anders Michelsen[1], Ib Friis[1] and *Lisanework Nigatu[2]*
[1]University of Copenhagen, Denmark and [2]University of Addis Ababa, Ethiopia

Are *Eucalyptus* plantations in Ethiopia ecologically harmful? This question is being addressed by a research group from the Biology Department, Addis Ababa University and University of Copenhagen. In this work, 100 forest stands from 10 plantation sites are being described floristically and edaphically in order to get a picture of the ecological effects of the widespread planting of *Eucalyptus* in various climatic zones in Ethiopia.

The results of such broad surveys are, however, not always easy to interpret, as stand characteristics such as litter and herb/grass coverage are seasonal. Consequently, permanent plots were established in selected 40 year old stands of the exotics *Cupressus lusitanica* and *Eucalyptus globulus* (coppiced) in a plantation of the indigenous *Juniperus procera* and in a natural *Olea-Juniperus-Podocarpus* forest in the central highland of Ethiopia, at an altitude 2400 m. Although the rainfall here is high (1900 mm per year), the site is fairly representative for large-scale plantations in Ethiopia, as these are most often found in the highland. The monthly litterfall of leaves and fine branches was followed over a period of two years together with the decomposition rate of litter, placed in nylon mesh bags on the forest floor, in order to quantify the nutrient element input to the forest floor through the litterfall and the subsequent nutrient release from litter to the soil through decomposition. Selected results from these investigations are shown in Figures 1 and 2.

The total annual litterfall in the four stands was 4.5, 5.6, 5.9 and 9.8 t/ha/yr in *Cupressus, Eucalyptus, Juniperus* plantations and in natural forest, respectively. It was generally high during the drier months and lower during the wet months (June–August), although the four stands showed different periods of peak litterfall (Fig.1). The litterfall in the natural forest was significantly higher than in the three plantations in 13 of 21 months studied so far, and it is in the upper part of the range recorded for moist tropical forests. The lower litter production in the three plantations may be due to the fact that most of the plantations are still in a rather early successional stage, in which most production is maintained in the live biomass.

The nutrient content of the litter from the four stands, as exemplified by nitrogen in Figure 2, showed marked differences between the tree species, in the following order: *Cupressus* < *Eucalyptus* < *Juniperus* < natural forest. As a result, less nutrients (nitrogen) are returned to the forest floor in the two plantations with the exotics *Eucalyptus globulus* and *Cupressus lusitanica*. The slowest decomposition was found in the natural forest, with 15% dry matter of litter remaining after 18 months, whereas 8% dry matter remained in the *Eucalyptus globulus* plantation. Consequently, and not surprisingly, the best protection of the soil against erosion is found in the natural forest, with a very high litterfall and a slower turnover of litter through decomposition processes.

With a knowledge of nutrient inputs with leaf and fine branch fall and together with the amounts of nutrients released from decomposing litter following litterfall in the four stands, an annual budget of important aspects of nutrient cycling in the sites can be established. From this budget it appears that the accession of nutrients to the sites for *Cupressus* and *Juniperus* is balanced with return, negative for *Eucalyptus* and positive for the natural forest. It thus appears that the *Eucalyptus* is depleting the soil of important elements in the investigated site. Export of nutrients from the sites with logging was not determined but would be expected to influence the total nutrient cycling considerably. This may especially be true for the fast-growing *Cupressus* and *Eucalyptus* species, from which the accession of important nutrient elements such as nitrogen, phosphorus and magnesium to the soil with litterfall is low; hence a major part of nutrients are found in the woody standing crop.

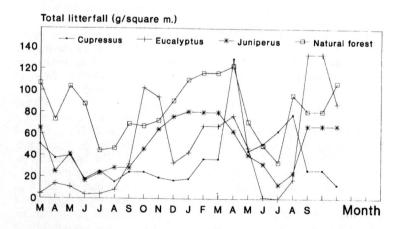

Figure 1. Litterfall in the four forests.

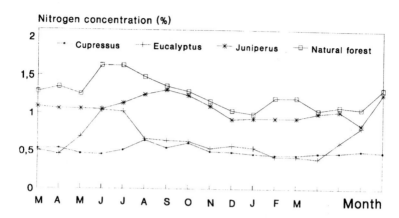

Figure 2. Nitrogen content of the litter.

The use of the exotic tree species, *Eucalyptus globulus* and *Cupressus lusitanica*, may thus not represent a sustainable land use on long terms, whereas the indigenous *Juniperus procera* in this respect may be recommended. Bioassays with local crops and trees as *Eragrostis tef*, *Chloris virgata* and *Acacia abyssinica* grown in the plantation soils support this conclusion, as these plants performed poorly in the soil of the *Cupressus lusitanica* and, notably, the *Eucalyptus globulus* plantation. The in-depth knowledge of nutrient cycling in these four forest stands will prove useful when the floristic and edaphic characteristics of all 100 forest stands investigated in the full range of climatic conditions in Ethiopia are evaluated.

Acknowledgement

This study was supported by the Danish Council for Development Research, Danida, grant no. Dan.8/481. The paper is extracts of an article which will be published by Lisanework Nigatu and Anders Michelsen entitled "Litterfall and nutrient release by decomposition in three plantations and a natural forest in the central highland of Ethiopia."

Agriculture and Sustainability in a Danish NGO Context

Henning Høgh Jensen
Danish Voluntary Service, Denmark

Introduction

Former volunteers from the Danish NGO *Danish Voluntary Service* (MS) of the agricultural sector were gathered in October to discuss the possibilities for ecological and economical sustainable agricultural production in a socially just way in the Third World.

The Brundtland report concluded that it was necessary to focus on the following strategies in order to achieve sustainable development:

1) an economic system which is able to generate surplus and know-how on a self-reliant and sustainable basis,

2) a production system which can search for and create new solutions to problems,

3) a technological system which promotes sustainable designs for business and finance,

4) a political system which secures people's participation in decision making,

5) a social system which gives solutions to conflicts arising as a result of disharmonious development,

6) an international system which promotes sustainable designs for business and finance and

7) an administrative system which is flexible and able to correct itself.

The first three strategies will only have a chance to work, if the last four are sufficiently organized to support the first three. The MS-workshop only worked on a few of these strategies, but we found it extremely important that MS attempts to conceptionalize the slogan sustainable development.

View points of practicians

What could this mixed group consisting of thirty practically-oriented participants contribute? Our common platform was our practical experiences from cropping systems in tropical conditions, particulary, with soil, husbandry, *etc.,* factors that are unique for local areas. The organizing committee hoped to put some flesh and blood on the theories by these practical experiences of the participants.

To make the concept workable, presentation-papers had been made in advance on five different themes: cropping systems, cultivation technology and erosion, extension and information, use of chemicals and pesticides and cycling of nutrients.

The workshop concluded that the understanding of the farmers way of thinking, their rationality was resonably good, because the rationality of farmers in the Western World is amazingly similar to that of farmers in the Third World. They have to consider the same elements, although farmers from Mozambique are not protected against price fluctuations as many are in the north. The complexity of the market system and vulnerability of the production system made it difficult to agree on a number of topics during the workshop.

Degrees of sustainability

To achieve ecological and economical sustainability one first has to acknowledge that the two may be mutually contradictory. This recognition is important, and everyone must make compromises to achieve the optimal sustainability. Naturally we hereby acknowledge that there are different degrees of sustainability.

We weighted the factors, we had to consider in a different manner. This reflected that optimal solutions for more sustainability will always be different from one place to another. Many complex factors interact demanding local knowledge and professionality of participants to respond to problems and propose appropriate solutions. We felt somewhat humble in front of this formidable task, but there was no simple solution.

Achievements from the MS-workshop

The following conclusions were among the most important achievements of the workshop: all models for solutions must build on thorough interdisciplinary professional cooperation, everybody recognizing their professional skills as well as their own limitations and dependence on others. The complexity of a sustainable cropping system complicates the introduction of new crops and systems, and it is evident that economical sustainability must be secured from the very start.

A good knowledge of local language is essential for development workers to be able to function properly in the local society. The language is the key to accept, cooperation, and interaction between donor and recipients in order to make the roles shift.

The participants had a rather pragmatic attitude to the use of pesticides in production, although we realize its many dangers and disadvantages. The elimination of the use of pesticides of purely ideological reasons is not resonable. On the contrary, a sustainable production is often only possible with use of pesticides. However, there is also a great potential awaiting exploration for use of the biological control of pests and weeds. Advantages and disadvantages of biological control must be considered in each case in relation to the need of increasing production, qualitatively as well as quantitatively.

Concerning the use of chemical fertilizers we found that the effect of social and economical dependency of the peasant family is a serious problem. Especially the unreliability of supply is critical, and we recommend the use of organic fertilizers wherever possible, although organic fertilizer will not always be an alternative to chemical ones. Chemical fertilizer should only be a supplement to organic fertilizer. We recommend that MS and their volunteers promote the use of compost, green fertilizers, mulching, legumes, inter-cropping, *etc.*, Further, there is a need for more knowledge on the effect of applying chemical fertilizers under different extreme conditions, *e.g.,* what is the effect on soil-acidity in semi-arid areas, and what is the effect on soil-acidity by growing legumes?

What is ahead?

It is important that MS faces the challenge to develop sustainability, both in ecological and economical terms, in a socially just way. Although it might seem contradictory, the policy of MS in the agricultural sector must promote a balanced weighting of these. The workshop recommends that MS concentrates its agricultural policy on the more marginal areas, *e.g.,* erosion affected, semi-arid and arid areas.

Considering the many uncertainties with different cropping systems the workshop, further emphasized the need for an organization, such as MS, to experiment! Designing experiments with the view of exposing, clarifying and conceptualizing the elements in a sustainable agricultural production. Thereby MS can clarify how development workers can contribute to promote sustainable development.

Famine and Colonial Administration

Jens Weise Olesen
Scandinavian Institute of African Studies, Sweeden

Introduction[1]

Since the Beja drought and famine of 1984 both national and international aid agencies have distributed food among the population of the Red Sea area in North Eastern Sudan. During the initial period of famine, food aid undoubtedly played an important role in saving human lives among the Beja. However, the Beja are now, according to several studies (*e.g.* Shingrai 1990) dependent on the continuous supply of food aid. Their own marketed food production suffers from a reduction in prices and the population is now even less able than in the early 1980s to supply their own community with locally produced food. Indigenous foods are being substituted by imported grain and a shift in the consumption pattern is taking place. The long-term impact of a continued distribution of free food is uncertain, but some reports point to a number of specific negative effects such as increased dependency of aid, migration to rural camps and centres, unemployment, break-down of social structures and corruption attached to the distribution of food aid.

The 1984 famine was precipitated by a severe drought. But this drought was not the first to hit the Beja. Droughts and eventually famines are not new in the Red Sea area. Through history various Beja tribes inhabiting north eastern Sudan (Red Sea and Kassala provinces) have had to cope with an almost unpredictable environmental and ecological situation. But they have always had a whole range of more or less effective socio-economic strategies and mechanisms at their disposal. The Beja have continuously been able to meet droughts and successfully survive harsh ecological conditions through use of alternative pastures and involvement in several productive activities like grain cultivation, camel porterage, trade, handicrafts, hunting and gathering of wild food, *etc.*

Historical records tell us that they have lived in the area for at least 3,000 years. And there is no indication that the climatic and ecological

---

1 This paper is based on a seminar lecture given by the author at the Scandinavian Institute of African Studies/Nordiska Afrikainstitutet, Uppsala, Sweden on January 31, 1991.

conditions have ever been fundamentally different from what they are today. The Beja have, especially during the last one hundred years, experienced prolonged and recurrent food shortages and also famines. During the Mahdist period (1881–1898) there was a serious famine, especially among the southern Hadendowa who, in the mid 1880s, supported the Mahdia (Paul 1971, Newbold 1935). Since then we find references to bad years and sometimes famines within short intervals.

It is, however, important to keep in mind that the Beja have been living "on the edge" for centuries. But several observers indicate that the Beja now face serious problems because their previously flexible and adaptive production systems are being fundamentally disrupted. The viability of Beja pastoralism is threatened because today the Beja system cannot counteract negative environmental changes.

The colonial as well as post-colonial state have neglected the pastoral production system of the Beja. No pastoral development projects have ever taken place in the area. Development efforts have been almost exclusively directed towards agricultural schemes and cash crop production. It is likely that the development of large scale irrigated cash crop schemes such as the Gash and Tokar Delta schemes in 1923 and 1895, respectively, had a negative effect on the pastoral production systems of especially the Hadendowa, Beni amar and Bisharin sub-tribes of the Beja, through a long term change in the inter-tribal organization of the pasture lands and removal of reserve pastures. The effects on Beja production systems of such schemes have not been subject to any historical analysis.

The deterioration of the environment and the pastoral economy of the Red Sea area have forced many Beja to leave their lands. They have migrated to the rapidly growing shanty towns, to Port Sudan or Kassala, for wage employment. The Gash and Tokar Delta schemes also attracted large numbers of pastoralists in search of regular wage labour or tenancy. Other sources of income are cash derived from the sale of locally produced charcoal for the growing populations in the towns and sales of livestock. Sale of livestock, especially camels are, however, a sign of desperate socio-economic dissolution because of their great social value to the Beja. Sale of these are considered to be a last resort before starvation. Restocking of for instance camel herds is extremely difficult because camels reproduce very slowly. Some Amarar of the Beja have not yet succeeded in rebuilding their camel herds after the 1947–49 famine (Dahl n.d.)!

The 1920 famine regulations

From the beginning of Condominium rule and up to 1920 there were local famines in the Sudan. In 1913 a serious, almost nation wide famine took place as a result of drought in most of the savanna belt. But at that time the Sudan government was much more concerned with the process of pacification than famine, and did not consider the approaching crisis as a major problem.

However, the local authorities in Dongola, northern Sudan, one of the most heavily affected areas, started their own relief measures. They distributed free food, established houses for the poor, organized relief works and intervened in the local grain market. In some places grain was sold at subsidized prices and some was distributed free.

There were also local famines in the Red Sea area on several occasions, but no serious government action took place before 1920, in spite of the fact that the major famine of 1913–1914 demonstrated the need for a coherent famine policy in the Sudan.

When the Sudan government finally decided to adopt a famine code for the whole country, it produced a set of regulations almost identical to the Indian famine code. The close link between Indian famine relief experience and the Sudanese regulations is evident, not only in the text of the regulations themselves but also in the many famine reports deposited in the National Record Office (NRO). Here we find many references to the Indian experiences and regulations.

In India, the history of British rule was closely related to famine and famine relief. At the onset, the Indian government was only handing out famine relief in return for work, but later regulations were based on acceptance of the principle that the government should do all it could to prevent suffering during food shortages. Reluctance to intervene in private trade and the market ·was replaced by regulatory famine interference.

In 1920 the Sudan government faithfully adopted the Indian regulations, without actually being aware that the Sudan did not possess the governmental apparatus and bureaucratic expertise to implement them. The 1920 Famine Regulations were thus designed without reference to the administrative and infrastructural realities of colonial Sudan. In India the grain-, livestock- and labour markets were far more integrated and the economy was more commercialized than that in the Sudan. Indian famines were often related to general unemployment and high food prices. Sudan did not have a general unemployment problem, in fact, the country was often faced with a shortage of labour, especially in the eastern and riverine areas.

The 1920 Famine Regulations did, however, provide the Sudan with a tool for government action for more than three decades and they remained in force until independence in 1956. The Famine Regulations put strong emphasis on the responsibility of the government to prevent distress and famine. The Sudan government had a decisive role not only in political matters but also in the overall economic development of the Sudan during the Condominium years. Much of the economic investments and infrastructural developments were initiated by government, which almost as a rule disapproved of private investments and foreign influence. In the introduction to the 1920 Famine Regulations it is said, "...it is the duty of the Government to offer to the necessitous the means of relief in times of famine....it is all important that precautionary measures such as the remission of taxation should be announced as early as possible to put heart into the people, and in case of actual distress relief should be given at the earliest possible moment. When people are on the verge of starvation a day or two's delay in giving relief may reduce them so much in condition that recovery is hopeless or protracted."

The famine relief system was based on five general principles: 1) a system of intelligence and early warning, 2) testworks, 3) relief works, 4) distribution of free food and 5) control of the grain market.

Famine relief in the Red Sea area

During the Condominium period the Beja suffered drought and eventually famine several times and the Beja famine in this century do have a historical record, especially after 1920. The documentation of famines are primarily found in the NRO in Khartoum which keeps records of almost any aspect of the British administration in Sudan. Unfortunately, the famine files of the Civil Secretary's Office do not cover the whole period in which the famine regulations were in force (1920–1956).

On at least 50 occasions the British 1920 Famine Regulations were put into force throughout Sudan. In such a vast country, adjustment to local conditions was a precondition for a successful relief programme. This implied a strong reliance on the local as well as the central government authorities.

By 1920 the Condominium administration was still a direct rule. On the lower levels the administration was often staffed by Egyptians while the British officials enjoyed almost total monopoly of the higher positions. Egyptian nationalism in the beginning of the 1920s, however, made the British anxious about its spread to the Sudan. Eventually, a crisis also emerged in the Sudan and culminated in 1924 when all Egyptian civil servants were expelled. At that same time a devolution of native administration and financial responsibilities took place. This was closely related to the British concern with Sudanese nationalism, financial affairs and the revival of Mahdism. Because the British feared a nationalist threat from an educated Sudanese elite, they were unwilling to let trained Sudanese replace the Egyptians. Instead they took to the Lugardian model of indirect rule or native administration. The Powers of Nomad Sheikhs Ordinance of 1922 marked the beginning of indirect rule in the Sudan. Tribal sheikhs were appointed in an attempt to re-establish their authority in tribal affairs. However, as a safeguard against corruption and abuse of power, all administrative cases were to be reviewed by the local District Commissioner (D.C).

Indirect rule or native administration was not, of course, carried out overnight. In the beginning tribal leaders acted in most cases as paid servants of the government as they had little power to influence local matters. The government, however, became concerned with native administration and even tried to create native authorities where none had existed. Practically, native administration originated in a situation of permanent shortage of trained staff within the government, but it was much cheaper to staff the administration with natives than with professional civil servants. However, it was never intended that Sudanese should assume the positions of the former Egyptian officials. The British simply considered it too risky to leave too much in the hands of the Sudanese!

Due to lack of personalities and training, indirect rule was in practice semi-indirectly up to 1927, when The Powers of Sheikhs Ordinance extended indirect rule to include the sedentary population. But administrative problems still existed. In the Red Sea area the Amarar, Bishareen and Beni amer Beja were grouped in one nomad administration under one single D.C., and the Hadendowa inconveniently occupied territories within both Kassala and Red Sea Province. Hence, administration of the Beja was often confusing. In addition some of the tribes (*e.g.*, Bishareen of the northern Red Sea area) preferred not to have any tribal leader (Nazir) at all. But the British officials were optimistic and one put it this way, "the Beja administration presented few difficulties precisely because it was so primitive, there was little to administer!" (Paul 1971).

However, tribal leadership was important for the implementation of the famine regulations. The administration of the largest Beja tribe, the Hadendowa, greatly improved by 1932, when a new Nazir of their own was appointed. The Civil Secretary was pleased to see local affairs handled by a tribal leader and not by an external official.

In general the native administration had its shortcomings, especially in relation to the implementation of government regulations. Even when the local tax-collector or un-elected deputy was respected, there would be problems of competence. This was the reason why the British opened up to a certain degree of Sudanization after the return to Anglo-Egyptian partnership in 1936. The Local Government Ordinance of 1937 was supposed to substitute Native Administration with Local Government and a professional civil service was to be developed. This process took time, however, but it slowly paved the way for a more effective use of the 1920 Famine Regulations in the rural areas.

The organization of famine relief and relief works demanded bureaucratic and technical skills, also at the local level. Projects needed to be identified, work-force and payments assessed, engineers recruited, accommodation arranged, food and water supplied, *etc.,* And often it had to be done within a very short period of time. Another recurring problem was the lack of trained staff to collect, codify and report relevant local information about rainfall, crop yields, *etc.*

The NRO-records on the 1925-1927 famine in the Red Sea area refer to the problems of implementation of relief work. On one occasion relief work on flood banks was actually put to an end because of difficulties with the local Sheikh. The governor laconically reported: "...he simply made a mess of things!" The regulations did not take into account the shortcomings of the Native Administration in relation to relief operations. It was designed for implementation by a professional bureaucracy. A bureaucracy, which did not exist in the Sudan in the 1920s. However, in spite of these problems the Famine Regulations became a tool of great value on the local level. A broad-minded interpretation of the regulations among governors and District Commissioners made local adaptations possible.

Early warning system. — The intelligence system (*de facto* famine warning system) of the Famine Regulations was closely related to the Native Administration system. Sheikhs faced with food shortages reported or complained to the D.C. about the condition of his people. The D.C. would then make an assessment of the requirements after a personal inspection of the area. Relief might then be forthcoming.

Often the system operated in a slow manner, mostly because of a mismatch between the official policy and the local realities. Sheikhs had no formal procedure for reporting and D.C.'s and governor's reports were sometimes difficult to interpret. But the system was not without value. It created an obligation for government to respond to famine and no governor or D.C. could ignore signs of distress and famine.

However, in a Red Sea area context the slow implementation did not have any serious implications on famine relief in practice. The Beja pastoralists, usually of low income, were high on assets, including livestock, stored grain, wild food, dom nuts and firewood as well as other alternative resources. They also had a social capital in the form of networks of obligations which linked them with kins. In addition there was a presumable ability to tolerate hunger. These factors often would protect them from sudden local destitution and therefore make the onset of a famine slow and compensate for slow famine relief operation. It was hence not, in the case of the Beja, always necessary for the government to respond quickly to signals of famine.

Relief work. — Relief work was one of the backbones of famine relief in the Red Sea area, but from time to time they were faced with the problems of recruitment. Relief works, also large scale, were common in India, but the Sudan did not have "work-famines" like India. The monthly reports of the D.C. often mention shortages of labour particularly in the Gash and Tokar Delta schemes. Much of the labour force in these schemes were actually migrant labourers from West Africa and central Sudan.

In order to make labour more mobile and solve the problem of labour shortages, cheap or free railway-tickets were issued to drought stricken Beja as part of famine relief. They were then able to go to the schemes for work. The workers were usually paid below average salary, but they also received payment in grain and coffee and food was handed out free of charge to those left behind, women, children and old people.

Control of grain market. — Another principle of famine relief was government control of the grain market. This was, however, a sensitive issue. Firstly, because the regulations did not actually include this interference and, secondly, because it was official British policy to support free trade, also in grain. But as we will see, attempts to control the grain trade was to become an important aspect of famine relief. This was also closely related to the distribution of free grain.

In 1926 the government was proud to announce that the people were fed without any active intervention in the grain market. This was very unusual, however, as coherent control policies were established and the government became deeply involved in the market for grain. The NRO files contain lots of evidence of imports and exports control, control of internal movements of grain, subsidized sales and issues of free grain and seed loans. Exports of durra were often banned and export of grain from famine stricken areas prohibited. The government also imported large quantities of grain from India. During the serious famine in 1941–1942 the government purchased grain for market stabilization and free issues.

Another frequent response to famine was to cooperate with traders in order to encourage local trade. In the marginal areas of the Sudan trade was suffering from neglect by the merchants. The markets were simply too distant and transport costs too high. Subsidized transports by train was therefore common in some areas.

The government had a very ambivalent attitude to the grain market. On the one hand ideology was inclined to the free market, but practice was often state intervention. Contradictions, pragmatism and local adaptions characterised British colonial famine relief in the Sudan.

As the second World War was approaching a Sudan Resources Board was set up in order to keep a close watch on the resources of the country. All imports and exports were subject to licensing and the government became heavily involved in both price and market control. There were restrictions to the amount each individual could buy as the aim was to prevent traders to buy in bulk for speculation.

The prices actually went down, but at the cost of diverting part of the grain supplies to the uncontrolled markets. The grain market in the Sudan was highly fragmented and that was a critical issue in food policy, especially among the Beja in the Red Sea Hills as they were never self-sufficient in grain and therefore always in need of cash to buy it. But the amount of cash they had varied greatly from year to year. Further, the Red Sea area was a relatively marginal area with high transport costs, and altogether this implied that traders were unwilling to bring grain to these markets. When the Beja did receive money from relief works or sale of animals, their experience was that marketed food only became more expensive because of the limited supply. Hence, they did not get more grain for their increased money, instead prices rose. On some occasions, however, the D.C. or the governor tried to encourage traders to increase the supply in Musmar, Sinkat and other places in order to assist the Beja to buy grain.

Conclusion

Before an attempt is made to draw any lessons from this presentation, I would like to stress that everything stated here has been based on written material, reports and correspondence deposited in the NRO, Khartoum and some local files in Sinkat in the Red Sea Province. All relevant material has not yet been consulted, but these sources, however, only represent one side of the story. The Beja themselves have not yet been consulted and oral histories collected among the Beja will probably provide us with important information about their own attitudes, perspectives and responses to British colonial famine relief.

What were the Beja perception of famine and famine relief? What did relief work mean to their socio-economic and socio-cultural system, their "way of life," and to the individual? Today we know that the Beja are not primarily concerned with individual survival, but to preserve for the future their pastoral way of living. And this was and is based on livestock plus some supplementary strategies such as labour-migration, sale of dom nuts, gathering of wild plants, hunting, charcoal production, camel-porterage, *etc.*, During the Condominum period they were faced with the demands and opportunities of a colonial state policy. The agricultural schemes (Tokar and Gash) took away grazing lands, but they also provided grain and grazing after harvest. The construction of the Port Sudan harbour also provided some Beja with cash obtained from wage labour, as did the urban economy of Port Sudan.

These developments affected the Beja in opposing ways. Drain of labour possibly put some constraints on their pastoral production system, especially as Beja pastoralism is labour intensive. On the other hand most Beja were dependent on cash and the market in order to obtain grain, *etc.*, They were never self sufficient with grain from their own flush-production. Only after some very good years had a surplus been registered locally.

Did colonial famine relief help the Beja to preserve their way of life? From the colonial source material one tends to think the answer is yes! During the 30 years the regulations were in force, many droughts and famines came to the Red Sea area and in certain periods people became destitute and some died. Many lost their livestock, but no famine like the great Sanasita disaster ever came to the Beja during this time.

The famines of the 1920s and at the beginning and end of the 1940s were serious, but they did not put a fundamental threat to the basic structures of Beja society and their way of life. However, people and animals sometimes died. In some cases Beja lost up to half of their camels and two thirds of their smaller livestock. And in a few cases the local D.C. and governor adopted a wait-and-see policy, like in the 1948–1949 famine, when the Hadendowa Beja asked for relief and the government expected that a good harvest in the Gash would be able to carry them through the coming season (CivSec 19/1/1 unpublished report). This was, however, an unusual response and grain was shortly after given in relief.

There are reasons to believe that without the British famine relief, many Beja would have become destitute during this century. But it can only be pure speculation, as to how the government would have responded to the great famine disasters before and. after the period in which the 1920 Famine Regulations were in operation (1920–1956). The regulations were applied almost 50 times during the period and without doubt they were instrumental in preventing major disasters. Major, nationwide famines required a massive intervention by the government and the famine regulations provided for this. The bureaucratic (and infrastructural) capacity to implement such a massive relief system existed only to a limited extent, but the system of local, pragmatic response, as was typical of the British famine relief in the Sudan through 30 years, did prevent serious local famines from developing into major disasters.

One could now ask — why this concern for preventing famine in colonial Sudan? Why not leave the marginal areas and their people to their own fate? The Beja were marginal and in many other respects the government neglected them and their pastoral production. There are several answers to these questions, both political and economic.

First, the British justified their presence in the Sudan partly on the grounds that a famine disaster like the Sanasita should never be tolerated. Furthermore, the Sudan was not a colony, it was, at least on paper, ruled as a Condominum by both Egypt and Britain. This, together with Egypt's claim on the Sudan, made it imperative for the British to justify their rule.

That political considerations were important in relation to famine relief, is very well documented in the Condominium files. Letters from governors and the Civil Secretary often refer to the public opinion both in Sudan as well as in Egypt and in Britain. Both British and Egyptian newspapers were concerned with famine issues in the Sudan, and the government was anxious not to release too much information about famine and failures to cope with it. In August 1932 famine killed 34

Nubas in Kordofan and the Civil Secretary reprimanded the governor in El Obeid for not having covered the case: "...it should not have been wired in clear. We don't want the press to make a story out of such a case (whether it is substantiated or not...)" (CivSec 19/1/1, unpub.).

The Egyptian newspapers, especially after independence in 1922, reported the Sudan famines and prominent Egyptians donated money to the victims. Famine relief was a political matter, and the legitimacy of the government was an important consideration in famine relief. Sudanese nationalists were able to mobilize the public on famine issues (in Kassala town 1948) and Egypt was making political advantages out of famine (Fung 1932).

To the British, the political advantages gained from distributing famine relief and organizing relief works were obvious. In 1932 one governor was pleased to report to the Civil Secretary that, "...they had realized for the first time that the government had other functions than to collect the annual poll tax. It is hoped that the value of relief work carried out will thus have a political significance" (CivSec 19/1/1, unpub.).

Famine relief was often preceded by tax or tribute remission. This took place on many occasions in the Red Sea area and the Sheikhs' response were of course positive. During my visit to the Red Sea Province, I was very often asked by Beja if I could bring back the British! "Our government does not care for us, but the British were good," many said. Now, in general the Beja and especially the Hadendowa Beja, with whom I spoke, have a very negative opinion on the government, and their history is characterized by recurrent droughts and famines. Very often they blame the government for being responsible of bad years. The recent Sudan government was as well being blamed for its neglect of the Beja people.

It is probably difficult to get a true picture of the Beja perception of the colonial regime as they tend to idealize the past. Daily life in the Red Sea Hills today is difficult for many Beja and often the past is remembered as "the good years" and the ones worth remembering.

Besides the political considerations one should also stress economic aspects. Labour shortages were a problem during long periods of the Condominium, not least in eastern Sudan, but in periods of severe drought and famine many Beja were forced to leave the Red Sea Hills in search of a livelihood, either in town or at the agricultural schemes. The famine regulations also provided for relief works and even though it was sometimes difficult to recruit enough labour, they usually were quite successful.

The regulations were to a large extent centred around public relief works, such as road construction, digging of hafirs (water reservoirs), repairing of railroads and irrigation channels and treeplanting. Usually the relief works were on a small scale, but on one occasion a large scale relief work took place in the Red Sea area. That was the building of military facilities in Port Sudan.

The small scale perspective brings us to another important aspect of famine relief in the Red Sea area. During the 1940s a large number of local relief works were implemented in and around the Red Sea Hills. These works were generally very successful and the reason for this was their small scale and local orientation. Drought-stricken Beja did not have to migrate out of their home area and leave their animals and families, and the local administrator was able to use his knowledge about local conditions.

The regulations were actually not drawn up for a local perspective, but local administrators very often became aware of this problem and they soon made their own local and unwritten variety of the regulations. When the government realized this they immediately emphasized that: "...the regulations should not be followed to the letter ... but they do lay down the principles ... and they form a useful corpus of advice of any administrative authority" (CivSec 19/1/2, unpub.). The need for a local approach in famine relief and adjustment to local conditions are some of the lessons to be drawn from the British famine policy during the Condominum.

The government even encouraged local organizations and traditional leaders to take part in the distribution of famine relief. And in some cases this encountered some difficulties as preferences for giving to kinsmen often existed. Some of this relief was even re-sold at the market. The British officials of course disliked this practice, but one must bear in mind that during a serious famine social cohesion is often threatened. And in times of crisis the Beja aimed at preserving their livestock and their social cohesion, not only the material well-being of the individual.

"The giving of food to rural people should not be seen as an nutritional support. Many rural people will sell part of the food they receive. Most relief agencies currently disapprove of such sales. However, famine relief can only be truly effective if the recipients are able to use the resources they are given as they see fit, and this may well include selling them" (de Waal 1989).

Finally, regarding the intelligence system (warning system), one specific lesson to be drawn from the experience with the famine regulations, is that the government of today will have to rehabilitate the civil service in the marginal rural areas and strengthen local participation in order to raise the quality of communication between the people and their rulers. Any intelligence system must be designed to fit in with existing local government and an understanding of local conditions.

Colonial files in the Sudan are an important source of information regarding early experience with food aid and famine relief programmes (Olesen 1990). This is especially so as public response to prevent famine disasters before and after the period in which the regulations were in force was insignificant. The British had more than 30 years of experience from the Sudan and any organization, agency or consultant

concerned with food aid and other relief measures would benefit greatly
from consulting the rich sources which exists in the NRO as well as the
local archives throughout Sudan. The country has a well documented
colonial history as the British kept records on almost any aspect of their
colonial administration.

Literature cited

Shingrai, A. A. 1990. The Impact of Food Distribution by the Aid Agencies During
the Famine Period (1984 onwards) in the Sinkat District. — Ahfad University for
Women, School of Family Sciences. Omdurman, Sudan.
CivSec 19/1/1 Unpublished. Famine, 1932–36. — National Records Office,
Khartoum, Sudan.
CivSec 19/1/2 Unpublished. Famine, 1937–49. — National Records Office,
Khartoum, Sudan.
Dahl, G. (n.d.). Who can be Blamed? Interpreting the Beja Drought. — Mimeo,
Department of Social Anthropolgy, University of Stockholm, Stockholm.
Newbold, D. 1935. The Beja Tribes of the Red Sea Hinterland. — in: Hamilton, J.
A. de C.(ed.), The Anglo-Egyptian Sudan from within London.
Olesen, J. W. 1990. Kan vi lære af historien? Refleksioner over arkivmateriale i
National Records Office, Khartoum, Sudan. — Nytt från Nordiska
Afrikainsitutet 27: 25–29.
Paul, A. 1971. A History of the Beja Tribes of the Sudan (2nd ed.). — Cambridge
University Press, London.
Waal, A. de 1989. Famine that Kills, Darfur, Sudan 1984–85. — Clarendon Press,
Oxford.

Notes from Concluding Session and Evaluation

Mike Speirs
Copenhagen, Denmark

Review and discussion of main topics

Professor K. K. Prah summarised the presentations and debates during the workshop, drawing attention to the salient and exciting points which had been made on many different issues. These included the problems of structural adjustment programmes in the Sahel, trade and world markets, land degradation, carrying capacity and land use management, as well as in project aid efforts. It was noted that working group discussions had focused on:

1) The need for a careful assessment of land tenure and land use management problems in the context of sustainable development, involving governments, project executing agencies and other organisations.

2) The need for multidisciplinary teams and different types of expertise in project planning and execution.

3) The wide ranging difficulties which are encountered in efforts to develop participatory approaches to rural development in the Sahel.

It was also noted that although the concept of sustainable development had been examined from various angles during the workshop, there was still a lot to be done in order to understand all the ramifications of this subject. Similarly, in the development of better economic policies, in the improvement of social conditions and in the reform of international trade, many different issues must be taken into account in future research.

During the subsequent (wide ranging) discussion, participants noted that development agencies in the Sahel must operate at many different levels in order to assist in the formulation of appropriate social, economic and ecological as well as political practices and policies. The need to continue the exploration of sustainable agricultural production systems in the Sahel, through a search for a common language on which future research and policy debate can be based, was emphasised. The difficulties which arise in the course of development project work were also underlined, and it was suggested that there is a need for greater humility and an acceptance of trial and error in the elaboration of development schemes and in the application of research results and ideas.

Suggestions for next workshop

The following topics were proposed, and will be considered by the new preparatory and organising committee for the workshop in 1993 (listed in no particular order, and without taking possible overlapping into account):

1) The development of education and agricultural extension services in the Sahel,

2) The dynamics of cash crop production (including gum Arabic),

3) Integrated management of natural resources, including land tenure and economic policy issues,

4) Political developments and the role of governments and politicians in the Sahel,

5) Decentralisation and institutions (government and local authorities) in land use management,

5) Fresh water and the development of water resources

6) "The enabling environment" (organisations and institutions),

7) Sustainability of agricultural production systems in the Sahel,

8) Follow up to the United Nations Conference on Environment and Development (UNCED, Rio de Janiero, June 1992) and

9) Assessment of the needs, capabilities and capacity of the Danish resource base for work in the Sahel.

Furthermore, it was suggested that the workshop should start with one or two keynote lectures, followed by group discussions on specific topics.

Committee (1992/93)

It was agreed that the new preparatory and organising committee will include the following persons:

Key institutional contact:

Leon Brimer, Institute of Pharmacology and Toxicology, KVL Bülowsvej 13, DK-1870 Frederiksberg C.

Other members:

Jens Dolin (Maisons Familiales Rurales), Tinggarden 13, DK-4681 Herfølge,

Christa Nedergaard Rasmussen, Danish Red Cross, Dag Hammerskjolds Alle, 2100 Copenhagen,

Marlene Mayer and **Lars Krogh**, Institute of Geography, Øster Voldgade 10, DK-1350 Copenhagen K and

Agnete Jørgensen, Dept. of Systematic Botany, Institute of Biological Sciences, Aarhus University, Nordlandsvej 68, DK-8240 Risskov.

The 91/92 organizing committee was thanked for its efforts and the proceedings were closed by professor Sofus Christiansen.

Summaries of Participants' Projects

The Sahel Workshop is a forum where experiences are exchanged and a network of people, who work in the Sahel and similar regions, is created. The participants of the workshop were asked to make a summary of their projects. The summaries are printed with only minor editorial changes.

Andersen, Inge
Keywords. — Soil degradation, remote sensing, Geographical Information System (GIS), resource management, nomads, Niger.
Abstract. — With soil degradation as a point of departure, and with sustainable development as a central theme, we wish to use remote sensing and GIS to describe and analyse the land use systems in an arid area in the northern Niger. The area is populated by both nomadic and sedentary Tuaregs. This will be followed up by a field trip to the area that will provide us with ground observations and interviews with the inhabitants of the area during spring 1992.
Partner. — Lone Mouritsen.

Brauer, Ole
Keywords. — Development, Mauritania, Burkina Faso, Ethiopia, Erithrea, Sudan.
Partners. — Lutheran World Federation, local churches.
Main activities of institution. — Emergency relief, development, develop-mental education.

Bregengaard, Per
Title. — Cooperation with the NGO Maisons Familiales Rurales (MFR) in Senegal.
Keywords. — Development patterns, fundraising, evaluation, education.
Abstract. — I have worked in the Danish organisation Maisons Familiales Rurales's Friends in Denmark with fundraising and evaluation of the supported projects. The organisation achieved some experience with smallscale development projects at the local level throughout Senegal. We have concentrated on agricultural projects with different amounts of education involved.
Partners. — Maisons Familiales Rurales in Senegal.
Main activities of institution. — Fundraising for financing projects in cooperation with local MFR's in Senegal. Evaluation of projects and discussion of project applications. Collection and registration of material about MFR and Senegal in general. Spreading the knowledge of MFR and other grass-root organisations in Senegal. Arranging personal contacts between local MFRs and Danish people. Importing and selling briefcases made of recycling materials.

Brimer, Leon
Main activities. — Polyphenolics in *Acacia* gum exudates, coumarin molluscicides, structure-activity relationships.
Keywords. — *Acacia*, polysaccharide, polyphenolics, tannins, exudates, coumarins, furanocoumarins, schistosomiasis, molluscicides.
Abstract. — Polysaccharide (gum) exudates may contain both monomeric and polymeric polyphenolics. These may possess significant anti-nutritional effects or may even be strongly physiologically active. Only few and scattered investigations have been

presented concerning polyphenolics in exudates from *Acacia* species. The aim of the present project is to investigate the polyphenolics in *Acacia* exudates and to anayse this survey with respect to (chemo)systematic implications within the genus. Natural as well as synthetic coumarins and furanocoumarins are tested for their molluscicidal activity in order to establish a quantitative structure-activity relationship for this interesting group of potencial molluscicides.

Partners. — Bruce Maslin, Dept. of Conservation and Land Management, Western Australian Herbarium.

Main activities of institution. — Insecticides/acaricides against ecto-parasitic diseases in domestic animals: uptake, residue concentrations, biological assays. Physiologically active, secondary plant constituents, structure function chemical analysis.

Brøgger-Jensen, Steffen

Title. — Study on Sahelian bio-diversity by means of satellite image analysis and ornithological surveys.

Keywords. — Sahel, habitat, satellite imagery, ornithological surveys, habitat selection, ornithological diversity.

Abstract. — The aim of the project is to develop a method for mapping and monitoring of biological diversity in the Sahel zone, permitting implementation of strategies for the conservation and monitoring of bio-diversity. The Sahelian biotopes hold a number of endemic biological assets and constitute the main wintering ground for a significant number of Palearctic migratory bird species. The relations between the distribution of bird species and communities and the habitat types and land-use patterns will be elucidated by means of satellite data analysis (CHIPS software, developed by the Inst. of Geography, University of Copenhagen) and standardized ornithological survey methods. On the basis of the habitat selection patterns revealed in the analysis, a framework will be established for the use of bird species and bird communities as biological indicators to set priorities for conservation of bio-diversity in large geographical areas. While carrying out ornithological surveys, this approach should at the same time ensure a further understanding of patterns and processes in the Sahelian ecosystems. Census areas have been placed in Senegal, where ground surveys for satellite image analyses and botanical surveys have been conducted, contributing with descriptive and analytical information on the environment.

Partners. — Centre de Suivi Ecologique, Dakar, Inst. of Geography, and Inst. of Population Biology, Copenhagen University.

Main activities of institution. — Nature management.

Carus, Hanne

Keywords. — Agricultural training and extension, natural resource management, community participation, environmental conservation, sand dune stabilization, Niger.

Abstract. — Environmental conservation activities are being carried out as an integrated part of a drinking water supply and small scale irrigation project in eastern Niger. In some of the project areas sand encroachment is a threat to drinking water infrastructures, agricultural areas and inhabited areas. In order to secure the physical sustainability of these areas and installations, conservation activities were included in the different project components. Community participation plays a central role in the programming and execution of all activities and different approaches were applied, depending on the nature of the individual activities. The project is concerned about different aspects of community participation in environmental conservation activities, such as 1) type of participation, on who's premises? 2) importance of the resources of the beneficiaries (physical, financial), 3) influence of the attitude and consciousness of the beneficiaries, 4) role of local institutions and organizations in the community, 5) project involvement/support, 6) choise of techniques, and 7) timing of activities.

Main activities of institution. — Country services outside Scandinavia within the fields of water supply, rural development and environment.

Degnbol, Tove
Title. — Evaluation of the Sudan-Sahel-Ethiopia programme 1986–1990 (an assignment for the Ministry of Foreign Affairs, Norway).
Keywords. — Research activities, NGO projects, multilateral agencies, food production, environmental protection, Mali, Ethiopia.
Abstract. — An evaluation of large programme to support food production and environmental protection in Sahel, Sudan and Ethiopia. The evaluation includes 1) assessment of the relevance of the support channeled through Norwegian NGOs, Norwegian and Sahelian, Sudanian and Ethiopian research institutions and multilateral agencies, 2) country studies on Mali, Burkina Faso, Sudan, Ethiopia and Erithrea (desk studies comparable to Danidas situation and perspective analyses), 3) field evaluation of selected projects in Mali, Ethiopia and Erithrea and 4) management study of the administration of the programme. The final report is expected to be submitted in June 1992.
Partners. — Dept. of International Development Studies, RUC and British and Norwegian researchers.
Main Activities of Institution. — Identification, appraisal, implementation, review and evaluation of projects in Denmark and abroad. Of particular relevance to the Sahel countries can be mentioned 1) preparation of situation and perspective analysis for DANIDA (Senegal, Niger, Guinea Conacry), 2) preparation of a report on Danish assistance to the CILSS countries 1980–1990, 3) implementation of an UNSO/DANIDA founded afforestation project in the Sudan, and 4) implementation and preparation of a number of water supply and electricity projects in the Sahel countries.

Dolin, Jens
Same as Bregengaard, Per

Frederiksen, Pia
Keywords. — Mapping and monitoring of soil degradation indicators by digital satellite image interpretation semi-arid lands, Landsat-MSS, soil degradation, Kenya.
Abstract. — The aim of the project is to contribute to development of methodologies for using satellite images to map information of degradation processes within the soil/vegetation complex in semi-arid lands. Moreover, to investigate the possibility of monitoring of changes over years in the surface condition. The emphasis has been laid on estimations of vegetation cover in a rangeland scene, which will inevitably consist of green as well as senescent vegetation. Minor investigations have included radiometric measurements of soil surface types.
Partners. — Dept. of Geography, University of Nairobi.
Main activities of institution. — Research and education within geographic fields including environmental geography.

Furu, Peter
Keywords. — Man-made lakes, irrigation, water resource management, primary health care, disease vector control, environmental management.
Abstract. — In its capacity as a WHO Collaborating Centre for Applied Medical Malacology and Schistosomiasis Control and Joint WHO/FAO/UNEP/UNCHS Collaborating Centre for Disease Vector Control in Sustainable Development, the Danish Bilharziasis Laboratory is involved in research, training and consultancies on environmental health issues in water resource development projects.
    The ever changing environment is to a great extent a result of human communities destroying their ecological support systems partly as a result of the population growth and the exploitation of global resources and ecosystems. To be able to meet the needs of the growing populations governments are forced to increase food production by development of new land (*e.g.* by deforestation) and intensive cultivation schemes and water resource projects. The resulting ecological changes in rural areas alter the disease patterns of a number of important vector-borne diseases such as malaria, schistosomiasis, onchocerciasis and Japanese encephalitis. The work is mainly on the health impacts of development projects such as irrigation schemes and environmental management. This is

the deliberate alteration of environmental factors or of the interaction between environmental and human factors. The proper operation and maintenance of agricultural schemes may be critical for the health status of the farmers.

The last 10 to 15 years have witnessed important efforts to promote the concept of environmental management for disease vector control. This implies that the planning organization carries out activities for the modification of environmental factors and their interaction with man to prevent or minimize vector propagation and to reduce man-vector-pathogen contact. These different approaches to a sustained disease vector control demands a profound knowledge of the epidemiology of the diseases as well as an understanding of the interrelationship between the biological, physical, socio-cultural and political environmental factors. This calls for an interdisciplinary approach and intersectoral collaboration.

Partners. — The WHO/FAO/UNEP/UNCHS Panel of Experts on Environmental Management for Vector Control and a number of collaborating institutions, mainly in Africa.

Main activities of institution. — The Danish Bilharziasis Laboratory (DBL) is a private institution, associated to the University of Copenhagen and sponsored by the Ministry of Foreign Affairs. DBL operates as a teaching, research, consultancy and service institution for water-related parasitic diseases in man and his domestic stock, with primary emphasis on African schistosomiasis, filariasis, dracunculiasis and malaria.

Grant, Stewart

Keywords. — Land management, forest management, water resource management.
Main activities of institution. — Provide consulting and management services to national and international organisations and private individuals in the field of natural resource management both in Denmark and abroad.

Higashidani, Ingeborg Kragegaard

Keywords. — Human settlements, camp-refugees, Somalia, gender relations.

Jacobsen, Jørgen Borch

Title. — Biomass briquetting in developing countries.
Abstract. — A World Bank pilot project on the use of agricultural waste as fuel is ongoing in Ethiopia with COWIconsult providing the consultancy services. The project comprises the establishment of five briquetting plants with a designed minimum output of 5,000 t of briquettes per annum. The plants are erected at state farms and the residues to be used comprise coffee husks, coffee parchment, cotton stalks, wheat straws, sugar bagasse and sugar cane tops.

Jensen, Dorrit Skårup

Keywords. — Socio-economic studies, rural development, water and sanitation, community participation.
Main activities of institution. — Country services outside Scandinavia within the fields of water supply, rural development and environment.

Krogh, Lars

Title. — Yield sustainability of the millet production in the Sahel.
Keywords. — Soil development, soil fertility, degradation, millet cultivation, sustainability.
Abstract. — An increasing awareness since the famines of the beginning of the seventies and environmental problems in the Sahel-region led to a discussion of agricultural sustainability of the Sahel. Desertification, deforestation and degradation are of increasing concern and there is an urgent need for research and education on farming systems that can increase productivity and profits without having adverse effects on the environment. The present project deals with the properties of soils in Burkina Faso and the aims to estimate wether any form of chemical or physical degradation has taken place and with the effects of continous cultivation on the nutrient balance.
Partners. — Anette Reenberg and Kjeld Rasmussen, Inst. of Geography, Copenhagen.

Lauridsen, Elmer B.

Keywords. — Seed procurement, tree improvement.
Abstract. — Research and development of improved techniques and methodology in seed handling and in evaluation of species and provenance trials. Implementation and evaluation of provenance trials in teak, *Gmelina arborea*, and (in collaboration with FAO) arid-zone species. Conduct of training cources at Danida financed tree-seed projects.
Partners. — Forestry institutions, projects in Danida´s priority countries, UNSO, FAO, etc.
Main Activities of Institution. — Seed procurement, gene resource conservation, tree improvement.

Lykke, Anne Mette

Title. — Sustainable use of natural vegetation in the Sine Saloum region in Senegal.
Keywords. — Naturel vegetation, Senegal, conservation.
Abstract. — The aim of my Ph.D. project is to analyse various aspects of the different vegetation types in the Delta du Saloum National Park, and on the basis of this to seek an understanding of the kind and degree of human impact possible without causing irreversible degradation of the vegetation.
Partners. — Institut des Sciences de l'Environnement (ISE); Université C.A.D. de Dakar, Senegal; Service des Parc Nacionaux, Senegal; and Institut Fondamental d'Afrique Noire (IFAN), Dakar, Senegal.

Markussen, Birgitte

Keywords. — Development, communication, educational video production.
Abstract. — The subject of my thesis for Magisterkonferens focuses on the use of video to motivate and educate rural populations. I have mainly been working in Latinamerica, especially in Mexico but have recently been 4 months in northern Nigeria as educational video production specialist in a development programme for Danagro Adviser A/S. The video productions made in Nigeria are on environmental issues.

Meyer, Marlene

Title. — Changes in agricultural land in Ghana.
Keywords. — Land-use, farming systems, remote sensing, Geographical Information System (GIS), vegetation degradation, change detection.
Abstract. — The aim of the project is to study the development in agricultural land-use in the savanna zone of Ghana in response to environmental and socio-economic factor, using SPOT-satellite images and GIS.
Partners. — Inst. of Geography and Remote Sensing Application Center, University of Ghana.

Michelsen, Anders

Title. — Ecological studies of fuelwood plantations in Ethiopia.
Keywords. — Survey of vegetation, root nodulation, mycorrhiza, soil, plant nutrition, revegetation, tree nursery, plantation.
Abstract. — The ecological effects of fuelwood plantations in Ethiopia is studied, with special emphasis on the effects of *Eucalyptus* on soil chemical properties and the vegetation of the forest floor. 100 stands of various plantation trees have been analysed in 10 sites in Ethiopia. Moreover, the nutrient cycling of four selected stands is followed intensively and the significance of symbiontic microorganisms as nodulating bacteria and mycorrhizal fungi is analysed in tree nurseries, plantations and natural forests in Ethiopia and Somalia. Simple techniques of inoculation in nurseries are currently being developed and applied.
Partners. — Biology Dept., Addis Ababa University, Botanical Museum, University of Copenhagen.
Main activities of institution. — The Institute is conducting research on ecological processes in terrestrial systems. The ecology of plants is studied on population as well as on ecosystem level. The emphasis is on interactions between plant, soil and microorganisms and the effects of man on the natural environment. The activities of the

126 *Lykke, Tybirk & Jørgensen*

Institute are concentrated in Denmark, but projects focusing on tropical regions are increasing in number, *e.g.* in Ethiopia, Galapagos, Reunion, Java and Somalia.

Lone Mouritsen
Same as Andersen, Inge

Nielsen, Jo Falk
Keywords. — Social change, Arabic languages, Islam, communication, information.
Main activities of Institution. — Education at B.A. and M.A. level in the Semitic languages and the cultures of the Semites.

Nøhr, Henning
Same as Brøgger-Jensen, Steffen

Olesen, Jens Weise
Title. — The Anglo-Egyptian administration and the Hadendowa (Beja), Red Sea area, northeastern Sudan, 1920-40.
Keywords. — Africa, colonial history, administration, pastoralism, state intervention, resource base, environment and economy.
Abstract. — A major argument in this study is that the basis for a long term change of the Hadendowa pastoral production system was layed during the 1920s and 1930s. A period which can be described as a period of regulations and large scale state intervention. Colonial state interventions during the inter-war period marked the hight of encroachment on the Hadendowa resource base. The study aims at identifying the long term consequences of colonial administration and intervention: pastoral decline, famine and increased vulnerable droughts. This process of intervention and integration was contradictionary: it increased the pressure on the Hadendowa resource base but it also provided the producers with new opportunities such as wage, labour, small scale industry, trade and migration.
Partners. — The study is a Dr. Artium project at the Dept. of History, University of Bergen. The research is carried out as part of the cooperation project between the University of Khartoum, the National Records Office, Khartoum and the University of Bergen.
Main Activities of Institution. — Scandinavian Institute of African Studies is an independent research institution, responsible to the Swedish Ministry of Foreign Affairs. The institute is financed by funds from Denmark, Finland, Norway and Sweden and is governed by a Nordic Research and Programme Council. It conducts research on Africa in the Nordic countries and it dissimates information about African affairs. The library is specializing in contemporary Africa. It organizes conferences, workshops and seminars and it has various programmes to support research on Africa: a travel- and study programme, a guest research programme and a publishing department. More than 300 titles have been published by the institute.

Petersen, Svend Kjeldgaard
Keywords. — Coordinating, supervision, research on projects.
Partners. — International Save the Children Alliance.
Main Activities of institution. — Project and emergency aid activities in Africa, Far East, Middle East, Eastern Europe, Latin America — with a special emphasis on children (and women) of groups in the worst situations socially and economically.

Rasmussen, Kjeld
Title. — Agricultural systems in northern Burkina Faso and satellite based monitoring of agriculture, vegetation and agro-climate in Senegal.
Keyword and Abstract. — Same as Anette Reenberg.
Partners. — Centre de Suivi Ecologique, Dakar, Senegal, CKTO, Ouagadougou, Burkina Faso.
Main Activities of Institution. — Same as Anette Reenberg.

Rasmussen, Christa Nedergaard
Title. — Bhutan — another development?
Abstract. — Modernity versus traditionality. Ecological and cultural sustainable development with focus on self-relevance and aspects of participation in a growth oriented development. Project administration and development projects in West Africa and Sudan. Especially environmental education for children in and outside the schoolsystem.

Rasmussen, Søren Skou
Title. — Afforestation and reforestation project in the northern region of the Sudan (UNSO/DANIDA).
Keywords. — Participation, land tenure, household interviews, shelterbelt.
Abstract. — The main task of the anthropologist is to asses the willingness and ability of villages in the project area to participate in afforestation and reforestation activities. Through village meetings, households interviews and studies on land tenure issue, it is decided wether or not villages that have applied for the project support are prepared to participate. Follow up visits to assess the social impact of afforestation activities and assistance to extension agents are other tasks.
Partners. — Forest National Corporation, The Sudan-Judanese Sociologists.

Reenberg, Anette
Title. — Yield sustainability in millet production systems in northern Burkina Faso.
Keywords. — Agricultural systems, land-use, nutrient circulation, satellite monitoring.
Abstract. — Sustainability and possibilities for expansion in the millet-production agricultural systems will be evaluated. The agricultural study is a continuation on earlier studies of Sahelian agricultural systems, mainly focusing on satellite based methods for monitoring land-use, production factors and yield. The new dimension is linking land-use to the nutrient circulation budget in the agricultural system and to the soil-water budget. Basic studies on the soil properties (chemical as well as physical) of the main soil types will be related to production estimates in order to point out important bottlenecks in the system.
Partners. — Kjeld Rasmussen, Ph.D. student Lars Krogh.
Main Activities of Institution. — Studies of agricultural systems and their resource basis, including climate and soils. Use of satellite data for agricultural and environmental monitoring.

Speirs, Mike
Title. — Studies of livestock development in the Sahel.
Keywords. — Animal resources, socio-economic development, economic policy.
Abstract. — Involved in investigating food security, agricultural development and economic policy in the Sahel through participation in three studies: 1) research into grain marketing, trade and food security in West Africa, based on work in Burkina Faso which focussed on cereals marketing, agricultural development and the economic policies of the revolutionary government from 1983 to 1987, 2) editorial work to complete a study of integrated farming systems in the Sahel carried out between 1986 and 1989 in Burkina Faso, Mali, Niger and Senegal which aimed to investigate the potential of, and the constraints on, the development of mixed (crop and animal) production and 3) consultant in a team investigating livestock development in the Liptako-Gourma region, including the presentation of specific project intervention proposals to the "Autorité du Liptako-Gourma" and the African Development Bank (1991).
Partners. — World Bank.
Main Activities of Institution. — Rural development consultants.

Søndergaard, Poul
Title. — Mediterranean dry zone afforestation.
Keywords. — Species diversity, adaptability, erosion control.

Vanclay, Jerry
Title. — Tropical rainforests.
Keywords. — Sustainable timber production.
Main Activities of Institution. — Temperate forestry and wood technology.

Zafyriadis, Jean-Pierre
Keywords. — Rural water supply, management, environmental planning.
Main activities of institution. — Consulting services outside Scandinavia within the fields of water supply, rural development and environment.

Address List of Participants

* *: presented an abstract in this volume*
Δ*: presented an article in this volume*

Andersen, Gitte, Odensegade 8,3, DK-2100 Copenhagen Ø, Denmark

* **Andersen, Inge,** Student of Geography and Communication, Roskilde University Centre, DK-4000 Roskilde, Denmark. Telephone +45 31815450.

Andersen, Kirsten Ewers, Rambøl & Hanneman, Bredevej 2, DK-2830 Virum, Denmark.

Arnesen, Odd, Scandinavian Instute of African Studies, P.O. Box 1703, S-75147 Uppsala, Sweden.

* **Brauer, Ole,** Project Officer, Danchurchaid, Skt. Peders Stræde 3, DK-1453 Copenhagen K, Denmark. Telephone +45 33152800, Telefax +45 33153860.

* **Bregengaard, Per,** Chairman, Maisons Familiales Rurales's Friends in Denmark, Wesselsgade 2, l.th, DK-2200 Copenhagen N, Denmark. Telephone +45 35373598, Telefax +45 35373598.

Δ **Breman, Henk,** CAPO, P.O. Box 14, 6700 AA Wageningen, the Netherlands.

* **Brimer, Leon,** Associate professor, Dept. of Pharmacology and Pathobiology, Royal Veterinary and Agricultural University, 13 Bülowsvej, DK-1870 Frederiksberg C, Denmark. Telephone: +45 35283162, Telefax: +45 31353514.

* **Brøgger-Jensen, Steffen,** Biologist, Ornis Consult Ltd., Vesterbrogade 140, DK-1620 Copenhagen V, Denmark. Telephone +45 31318464, Telefax +45 31247599.

* **Carus, Hanne,** Agronomist, I. Krüger Consult A/S, Gladsaxevej 363, DK-2860 Søborg, Denmark. Telephone +45 39690222, Telefax +45 39690806.

Δ **Cheru, Fantu,** the American University, Washington, D.C., U.S.A.

Δ **Christiansen, Sofus,** Professor, Inst. of Geography, University of Copenhagen, Østervold 10, DK-1350 Copenhagen K, Denmark.

Christensen, Thyge, Overdrevet 21, DK-8382 Hinnerup, Denmark.

* **Degnbol, Tove,** Consultant, COWIconsult, Parallelvej 15, DK-2800 Lyngby, Denmark. Telephone +45 45972211, Telefax +45 45972212, Telex 33 580 COWI DK.

* **Dolin, Jens.** Highschool lecturer in geography, Tinggården 13, DK-4681 Herfølge, Denmark. Tel phone +45 53676036.

Enevoldsen, Thyge, Inst. of Int. Dev. Studies, Roskilde University Centre, Postbox 260, DK-4000 Roskilde, Denmark. Telephone +45 46757711.

Frederiksen, Peter, Inst. of Geography, Roskilde University Centre, Postbox 260, DK-4000 Roskilde, Denmark. Telephone +45 46757711.

* **Frederiksen, Pia,** Hus 19.2. Box 260, Roskilde University Centre, DK-4000 Roskilde, Denmark. Telephone +45 46757711 (2462), Telefax +45 46754240, E-mail piafre gorm.ruc.dk.

* **Furu, Peter,** Environmental Health Biologist, Assistant Professor, Danish Bilharziasis Laboratory, Jægersborg Allé 1D, DK-2920 Charlottenlund, Denmark. Telephone +45 31626168, Telefax +45 31626121.

* **Grant, Stewart,** Project Coordinator (Overseas Department), Danish Land Development Service, Klostermarken 12, DK-8800 Viborg, Denmark. Telephone +45 86676111 and +45 86671293.

Hansen, Mogens Buch, Inst. of Int. Dev. Studies, Roskilde University Centre, Postbox 260, DK-4000 Roskilde, Denmark. Telephone +45 46757711 (2462), Telefax +45 46754240, E-mail piafre gorm.ruc.dk.

* **Higashidani, Ingeborg Kragegaard,** Ph.D. student, Dept. of Ethnography and Social Anthropology, Aarhus University, Moesgaard, 8270 Højbjerg, Denmark. Telephone 86272433 (272), Telefax 86272378.

Hvidt, Nanna, DANIDA, Asiatisk Plads 2, DK-1448 Copenhagen, Denmark.

* **Jacobsen, Jørgen Borch,** Project Manager, COWIconsult A/S, Parallelvej 15, DK-2800 Lyngby, Denmark. Telephone +45 45972211, Telefax +45 45972212.

Δ **Jensen, Henning Høgh,** Amalie Skrams Alle 9, DK- 2500 Valby, Denmark.

* **Jensen, Dorrit Skårup,** Socio-Economist, I. Krüger Consult A/S Gladsaxevej 363, DK-2860 Søborg, Denmark. Telephone +45 39690222, Telefax +45 39690806.

Junkov, Mette, Rambøll & Hannemann, Bredevej 2, DK-2830 Virum, Denmark.

* **Jørgensen, Agnete,** Student, Biological Inst., Dept. of Systematic Botany, Aarhus University, Nordlandsvej 68, DK-8240 Risskov, Denmark. Telephone +45 86210677, Telefax +45 86211891.

Kamerud, André, c/o Just Gjessing, Geographic Institutt, Oslo University, Blendern, Oslo, Norway.

Kjær, Inge Langberg, Vesterbrogade 80,3, DK-1620 Copenhagen V, Denmark.

Kristensen, Margit, Friendship Ass. Denmark-Burkina Faso, Kielshøj 78, DK-3520 Farum, Denmark.

* **Krogh, Lars,** Ph.D.-student, Inst. of Geography, Øster Voldgade 10, DK-1350 Copenhagen K, Denmark. Telephone +45 33132105, Telefax +45 33148105.

* **Lauridsen, Elmer B.,** M.Sc. (forestry), DANIDA Forest Seed Center, Krogerupvej 3a, DK-3050 Humlebæk, Denmark. Telephone +45 42190500, Telefax +45 49160258.

Δ **Leth-Nissen, Søren,** c/o Provincial Fish Culturist, P.O. Box 510738, Chipata, Zambia.

Lundberg, Hans Jørgen, Danida, Asiatisk Plads 2, DK-1448 Copenhagen K, Denmark.

* **Lykke, Anne Mette,** Ph.D.-student, Biological Institute, Dept. of Systematic Botany, Aarhus University, Nordlandsvej 68, DK-8240 Risskov, Denmark. Telephone +45 86210677, Telefax +45 86211891

Markussen, Henrik Secher, Inst. of Geography, Roskilde University Centre, Postbox 260, DK-4000 Roskilde, Denmark.

* **Markussen, Birgitte,** Stud. Mag., Dept. of Social Anthropology, Aarhus University, Moesgard, DK-8270 Højbjerg, Denmark. Telephone +45 86272433.

* **Meyer, Marlene,** Ph.D.-student, Inst. of Geography, University of Copenhagen, Øster Voldgade 10, DK-1353 Copenhagen. K, Denmark. Telephone +45 33132105, Telefax +45 33148105.

* Δ **Michelsen, Anders,** Research Fellow, Inst. of Plant Ecology, Uni-versity of Copenhagen, Øster Farimagsgade 2D, DK-1353 Copenhagen K, Denmark. Telephone +45 33324346, Telefax +45 33145719.

* **Mouritsen, Lone,** Student of Geography and International Development studies, Roskilde University Centre, DK-4000 Roskilde, Denmark. Telephone +45 31834936.

Mølgård, Kika, Danida, Asiatisk Plads 2, DK-1448 Copenhagen K, Denmark.

Nielsen, Henrik, Vesterbrogade 21, 2TV, 1620 København, Denmark

* **Nielsen, Jo Falk,** Cand. phil. in ethnography. Student in arabic, Institute of Semitic Philology. Rundhøj allé 34, 2tv., DK-8270 Højbjerg, Denmark. Telephone +45 86140024.

Nielsen, Ivan, Biological Inst., Dept. of Systematic Botany, Aarhus University, Nordlandsvej 68, DK-8240 Risskov, Denmark. Telephone +45 86210677, Telefax +45 86211891.

Nielsen, Mogens Laumand, DANIDA, Asiatis Plads 2, DK- 1448 Copenhagen K, Denmark.

Noppen, Dolf, Nordic Consulting Group, Jaris Mosevej 9, DK-2670 Greve, Denmark.

* **Nøhr, Henning,** Director, Biologist, Ornis Consult Ltd., Vesterbrogade 140, DK-1620 Copenhagen V, Denmark. Telephone +45 31318464, Telefax +45 31247599.

* Δ **Olesen, Jens Weise,** Research Fellow, Historian, Scandinavian Inst. of African Studies, P.O. Box 1703, S-751 47 Uppsala, Sweden. Telephone +46 18155480, Telefax. +46 18 695629.

Oksen, Peter, Inst. of Geography, Øster Voldgade 10, DK-1350 Copenhagen. K, Denmark. Telephone +45 33132105, Telefax. +45 33148105.

* **Petersen, Svend Kjeldgaard,** Programme Officer, Red Barnet, Brogaardsvænget 4, DK-2820 Gentofte, Denmark. Telephone +45 31680888, Telefax. +45 31680510.

Δ **Prah, Kwesi Kwaa,** Cape Town, P.O.Box 369, Constaintia, 7800 Cape Town, South Africa

* Δ **Rasmussen, Kjeld,** Associate professor, Inst. of Geography, Øster Voldgade 10, DK-1350 Copenhagen K, Denmark. Telephone +45 33132105, Telefax. +4533148105.

* **Rasmussen, Christa Nedergaard,** Danish Red Cross, Dag Hammer-skjolds Alle, DK-2100 Copenhagen Ø, Denmark. Telephone +45 31381444.

* Δ **Anette Reenberg,** Associate professor, Inst. of Geography, Øster Voldgade 10 DK-1350 Copenhagen. K, Denmark. Telephone +45 33132105, Telefax. +45 33148105.

Δ **Seck, Moussa,** ENDA, B.P. 3370 Dakar, Senegal.

Δ **Sihm, Poul,** PA Consult, Døvlingvej 2, DK-6933 Kibæk, Denmark.

*Δ **Speirs, Mike,** Danagro Adviser, Granskoven 8, DK-2600 Glostrup, Denmark. Telephone +45 43434590, Telefax. +45 43434049.

* **Søndergaard, Poul,** Associate professor, Kirkegaardsvej 3A, DK-2970 Hørsholm, Denmark. Telephone +45 42860641, Telefax. +45 42860774.

Strømgård, Peter, DANIDA, Asiatisk Plads 2, DK-1448 Copenhagen K, Denmark.

Svien, Egil, Nordic Consulting Group, Jaris Mosevej 9, DK-2670 Greve, Denmark.

Tybirk, Knud, Biological Inst., Dept. of Systematic Botany, Aarhus University present address c/o FAO, Apartado 1721-0190, Quito, Ecuador.

* **Vanclay, Jerry,** Professor, Dept. of Economics and Natural Resources, Royal Veterinary and Agricultural University, Thorvalsensvej 57, DK-1871 Frederiksberg C, Denmark. Telephone +45 35282225, Telefax. +45 31357833.

* **Zafyriadis, Jean-Pierre,** Managing Director, hydrogeologist, I. Krüger Consult A/S, Gladsaxevej 363, DK-2860 Søborg, Denmark. Telephone +45 39690222, Telefax +45 39690806.

Zethner, Ole, Danagro Advisers, Granskoven 8, DK-2600 Glostrup, Denmark.